PROCESS CONTROL COMPUTER SYSTEMS
Guide for Managers

Edited by
TOM G. STIRE

ANN ARBOR SCIENCE
THE BUTTERWORTH GROUP

Copyright © 1983 by Ann Arbor Science Publishers
230 Collingwood, P.O. Box 1425, Ann Arbor, Michigan 48106

Library of Congress Catalog Card Number 82-70705
ISBN 0-250-40488-5

Manufactured in the United States of America
All Rights Reserved

Butterworths, Ltd., Borough Green, Sevenoaks
Kent TN15 8PH, England

PREFACE

Many excellent books have been written in recent years that thoroughly describe the technical aspects of computer-based process control systems. Although we have not done an exhaustive search, we would feel comfortable in stating that all topics, from instrumentation to hardware to computer software, have been addressed exhaustively in one work or another. These volumes clearly and precisely tell the reader how to design and implement a computer control system, and many more undoubtedly will be written as the technology changes and new techniques are developed.

However, one key element has been either omitted from, or only alluded to, by works to date: the management of control systems, which is absolutely critical. Without proper guidance, such efforts are often destined to failure, or at best, likely to fall far short of everyone's expectations. However, the modern manager is often unaware of the technical aspects of the ever-changing field and must rely on others for the detailed work. Therefore, books that describe the "how-to's" of control systems, while very valuable to the practitioners, do not tell the nontechnical manager the key element needed to know what must be done.

This book approaches computer-based control systems from the manager's point of view. It does not tell someone *how* to do something; rather, it describes *what* must be done and *when* the activity should occur. It focuses on the elements of a successful project, leads the reader from step to step and presents the "do's and don'ts" associated with such an effort.

The first four chapters are tutorial in nature. They highlight various aspects of control systems, including hardware (Chapter 3), software (Chapter 4), and the differences between data processing and process control (Chapter 2). Chapter 5 discusses the benefits of digital control. Concepts presented there should be very useful to the manager faced with evaluating the cost effectiveness of computer control.

Chapter 6 is the "what not to do" chapter. All managers should read it before starting into a project so that they are aware of the pitfalls and how to recognize them.

Chapters 7 through 11 define five success steps to computer control system projects: planning (Chapter 7), detailed design (Chapter 8), implementation (Chapter 9), staffing (Chapter 10), and management and optimization (Chapter 11). Each discusses what should be done, the importance of each and the techniques that should be applied.

Tom G. Stire

ACKNOWLEDGMENTS

I would like to acknowledge the continued support and encouragement of EMA, Inc. in producing this work. Although it was not a company project, and all of the contributors worked on their own, it was the EMA philosophy that provided the common thread. Furthermore, without the company's support in terms of clerical assistance, as well as its frequent encouragement, the work would not have been possible.

I would also like to thank several individuals for their assistance. Of course, the contributors are to be thanked for their many evening and weekend hours, as are their families, who had to give up that time from the personal activities of family life. Also instrumental were Mike Novotny, Pat Maher and Roxane Sabean, who provided excellent artwork and technical coordination. This book could not have been completed without the enormous effort of Cheryl Karr. Her many, many hours of typing, proofreading and meticulous attention to the details of style and format were absolutely essential to the effort of consolidating ten different authors into one work.

Tom G. Stire is Vice President of Control Systems and Services for EMA, Inc., a St. Paul–based engineering firm specializing in the application of instrumentation and computers to process control, and in the data-processing needs of plant manufacturing and maintenance. EMA provides services in all phases of a computer-based system, from conception and the attendent planning phases, through design and implementation, to training, optimization and staffing assistance. Under Mr. Stire's direction, a staff of more than 25 apply the concepts presented in this volume to control problems throughout the country.

Before joining EMA, Mr. Stire held several technical positions related to process control. He spent six years with the Pillsbury Company, involved in the development of real-time process monitoring and control systems, and with the use of computer modeling, simulation and statistical data analysis of numerous manufacturing processes. He also held several positions with the Control Data Corporation, where he was lead programmer on several large computer control systems. This experience includes systems programs, and application software for direct digital control, supervisory control and management information reporting. Mr. Stire also has experience in data-processing applications, having served as a systems analyst for two years in the U.S. Army and, later, as Director of Systems and Programming for a national firm in Chicago.

Mr. Stire holds a BS (Business Administration) from the University of Minnesota, where he majored in Quantitative Analysis and Mathematics. He also completed graduate studies in Operations Research and Computer Science at the same institution.

CONTENTS

1. The Management Overview 1
   T. G. Stire

2. Process Control Is Not Data Processing 19
   D. I. Knudsen

3. Control System Hardware 41
   P. V. Manns

4. Control System Software 69
   T. M. Brueck

5. Benefits of Digital Control 87
   A. W. Manning

6. *Caveat Emptor* — Let the Buyer Beware 113
   R. G. Skrentner

7. Success Step 1 — Planning 145
   R. Graupmann

8. Success Step 2 — Detailed Design 167
   C. N. Williams

9. Success Step 3 — Implementation and Installation 207
   T. P. McConville

10. Success Step 4 — Staffing 229
    G. L. Bennett

11. Success Step 5—Management of Process Control and Process Optimization 253
    R. G. Skrentner

Glossary 275

Index 291

CHAPTER 1

THE MANAGEMENT OVERVIEW

Tom G. Stire

INTRODUCTION

Without exception, the technology of digital computers has grown faster in the past 20 years than any other technology known to mankind. This growth has been so dramatic that it is beyond most of us to grasp it or understand its significance. It has been said that if the automobile industry had experienced the same technological growth, cars would be able to travel 250,000 miles per gallon of gas and cruise at 500,000 miles per hour. Also, we would find it cheaper to throw away our Rolls Royce rather than pay for parking.

However, this phenomenal growth in technology has placed a significant burden on users of digital computers. Computer manufacturers continually develop products that increase capabilities and capacities. Within a very short timeframe, they introduce new and innovative products, often to replace products that users have yet to learn about and that certainly have not yet been applied in any broad sense. This is particularly true in the area of the emerging microcomputer technology, where new "parts" are developed seemingly overnight.

As difficult as it may be for the average technical person to keep up, it is even more difficult for the manager, who must make decisions and manage efforts that utilize this ever-changing technology. Even with a good technical background, this person can be "obsolete" in a matter of months and must rely on the technical staff or outside consultants to chart the specific course.

Consequently, there must be a framework in which the manager can function that will allow rational decisions and meaningful direction. This book deals with such a framework for the application of digital computers to process control and automation.

This framework, or methodology, has been utilized by the authors on numerous control system projects. It provides an excellent mechanism for the successful application of digital computers. While it is not the only way to install a working system, it does minimize the usual pitfalls of computer system application. If diligently followed, it will maximize success in a minimum of time.

The methodology discussed in this book focuses primarily on the large, plantwide computer system because the larger projects have the most significant and complex problems. However, the manager should not view the method as limited to those types of applications, but rather should look closely at the concepts being presented and select the details that apply to the project at hand. These concepts then should be scaled down (or up) to meet the specific situation.

The manager responsible for directing a small process control installation should not allow any of the steps to be skipped or "short circuited" merely because the application is small. Undoubtedly, there will be a tendency to do so, particularly on the part of the technical staff, who feel that the problem is too small to require any rigorous definition. However, close inspection will reveal many details that have not been considered and that, if left undefined, eventually will result in undesirable compromises and unnecessary delays.

Such thinking is especially common when the system is being planned. We frequently find control equipment that was purchased based only on concepts of what it was to accomplish. The details were left for later, only to find that the equipment purchased was inappropriate, poorly sized or perhaps totally unusable. *There is no substitute for detailed planning,* regardless of the system size. Details must be thought out at some point. It is the prudent manager's job to be certain that this is done, done properly and done at the right time.

OVERVIEW OF COMPUTER CONTROL

The other chapters of this book deal with the many facets of digital control systems. Each was written by a different author whose objective was to discuss fully a given aspect of computer control and to provide a reference chapter on a given topic for the manager to consult at a later date. While each chapter presents a different part of the "framework," each contains information relevant to the subject being discussed and may, therefore, contain some overlap with other chapters.

Process Control Versus Data Processing

An often misunderstood concept is that any given computer can be used for any given application. While this is technically the case for the most part, nothing could be farther from the truth in terms of efficiency, ease of application and time required.

One very commonly encountered misunderstanding of this concept is the distinction between "data processing" and "process control." It is not unusual for a manager to reason that the new computer controlling a process also could be used to do payroll or other similar functions. After all, a computer is a computer is a computer! However, the manager who uses this reasoning is overlooking several very significant differences including hardware, software, people and the importance of the computer to the process needs. While it may be possible to mix the two, the manager must carefully analyze the differences and decide whether the additional costs and potential compromises are justified. After all, a Piper Cub and an F-1 fighter are both airplanes but are significantly different in use, cost and applicability. Chapter 2 discusses the differences between data processing and process control.

Control System Components

Chapters 3 and 4 are tutorial in nature. They are intended for the manager with little or no knowledge of computers. They form the basis for understanding the components of a digital control system and should be understood before reading the chapters that follow.

The typical manager frequently considers "the system" to be only the computer itself or, at most, only the equipment in the computer room, if we are speaking of a larger control system. However, there is much more involved. The "control system" consists of hardware, software and people.

Hardware

Control system hardware includes everything from the operator interface [video display terminal (VDT), panel] through the end element in the process (instrument, motor, etc.). All components must be considered in all phases of the project because the result must be an integrated system. If any piece is dealt with to the exclusion of the others, the final integration, which is of course essential, may not be successful. Chapter 3 deals with the equipment and briefly discusses each component.

4 PROCESS CONTROL COMPUTER SYSTEMS

Software

While the equipment components are important to the control system, the computer programs, or "software," are of equal significance. Computer equipment cannot function without software and, although it is much more difficult to grasp because it cannot be felt or seen, it is extremely important.

To a certain extent, software is more significant because of its exactness. The hardware, similar to the Piper Cub mentioned earlier, is general in application. Certainly there is some customization involved in hardware, just as there are certain options that are ordered with the Piper. However, what makes the plane unique is the pilot, the flight plan, the cargo, etc. What makes the control system unique to an application is the software, and the manager must be certain that a great deal of attention is paid to its definition and implementation.

There are several categories of software when considering a control system. Chapter 4 discusses these in detail. It also considers management of software projects, a potential nightmare for the manager unfamiliar with the concepts and pitfalls.

People

People will make the control system a success. The specific people, depending on the project, will include the system manager, operators, maintenance personnel, engineers and programmers. These people must be acquired and trained (Chapter 10) and, above all, managed (Chapter 11).

Benefits

The application of digital technology to industrial processes is relatively recent. The first systems were installed in the late 1960s [1]. However, early systems were large, expensive and, more importantly, difficult to justify. In fact, many early installations were not as successful as their proponents had promised. For many companies, the resulting false starts did set digital process control back many years.

Currently, however, computer costs are a small fraction of what they were ten years ago. Software advances have greatly reduced the amount of time and professional programming necessary to implement computer control. Technological advances in such areas as data communications, signal multiplexing and storage media have made distributed control

THE MANAGEMENT OVERVIEW

networks possible. In short, digital control is significantly more cost-justified than in the past.

Chapter 5 addresses the benefits that can be realized. It deals with the tangible cost savings that are possible, depending on the specific application and the manner in which control is approached. It also presents a myriad of intangible benefits that, although they usually cannot be quantified, nevertheless will be realized if a digital approach is taken and if the project is properly designed and implemented.

Potential Pitfalls

Although this book deals primarily with the positive aspects of digital control, we would be remiss if we did not alert the manager to the common problems that so often cause computer system failures. Chapter 6 presents several such situations. Its intention is not to frighten the potential user, but rather to point out the things that *could* go wrong if one is not careful.

The problems discussed represent a range of problems. On the one hand is the manager whose staff is convinced that the system is very straightforward. This extreme typically purchases hardware without much thought to exactly what is to be done until the hardware arrives and must be applied. Also in this extreme is the user who decides to let a vendor create the solution to a vaguely defined problem. In both cases, the result often is a misapplication because of too little detail.

The other extreme is the "turnkey system myth," or the manager who defines the application well but who is not prepared to commit the necessary resources to support the system when the supplier is finished. All control systems, from simple analog control to a plantwide digital control, require ongoing support. They are always dynamic, ever-changing and always in need of management. A system can be carefully planned, meticulously designed and competently implemented, but if a structure is not in place to manage the result, it will be less than the success anticipated.

Steps to Success

All these pitfalls can be eliminated, or at least reduced, by the manager who follows the success steps described in Chapters 7 through 11. All five steps are required on all projects, although the size and complexity of the application certainly will dictate the relative amount of effort required.

6 PROCESS CONTROL COMPUTER SYSTEMS

Success Step 1 – Planning (Chapter 7)

Before any work is done on any system, the prudent manager will assure that the requirements of the system are carefully analyzed and documented. This entails a careful assessment of the needs and objectives to be met, the alternative approaches available, the criteria for evaluating the alternatives and a complete description of the selected solution. These details must be well thought out and then outlined in writing so that nothing is overlooked.

This step also will form the basis for the justification of the application. One of the evaluation criteria must certainly be economics because an acceptable return on investment must be realized. Careful definition of the needs and objectives will help quantify exactly what the system can achieve. Knowing the realistic results to be expected, the specific dollar savings can be quantified.

The manager must not overlook the significance of this step. It is important that expectations be defined clearly and agreed to by all involved before starting the project. It may be easier to assume that everyone knows what will be done and what the results will be; however, this type of project will change over time as the technical people encounter problems and are forced to make compromises. Similarly, some people have surprisingly short memories, particularly when the project takes months or years to accomplish. All in all, it is best to take a brief amount of time to document the plan.

Success Step 2 – Design (Chapter 8)

Once the plan is made and documented, the detailed design can begin. This process differs from the planning process because it deals with exactly *how* the system will be implemented, while Step 1 only addresses *what* the system would do. For example, Step 1 would define what process or equipment was to be controlled. It might even define it in terms of expected ranges of the controlled variables. However, there would not be enough detail to "build" the system. Step 2 produces these details of the design, including specific inputs and outputs, interlocks, control logic, contingencies and operator interface, among others.

While this step is one of the more obvious to most managers, it is also one of the most frequently neglected phases of a computer project. This neglect may be caused by lack of familiarity with the technology or by the apparent ease of changing later, when the software is modifiable. It certainly is true when dealing with relatively small or simple applications. For whatever the reason, it is too often oversimplified.

This is particularly true with respect to the software of the project. Detailed definition of the control programs is often overlooked completely, at least until it is time to install the equipment. However, the computer program does only what it is told to do—nothing more, nothing less. Every detail must be considered and every contingency planned for at some point, and the best time is early in the project. Again, the prudent manager will insist on adequate documentation of this step.

One last point deserves mention. If the system must be programmed by a "programmer" rather than by a "process engineer," this step takes on added significance. The manager must find some way for the process people to define *exactly* what they want done in a manner that can be readily understood by the programmer, who may be totally unfamiliar with the application. If this is not addressed adequately early on, the implementation phase can be long, expensive and fraught with misunderstandings.

Success Step 3 — Implementation (Chapter 9)

After the entire project is designed, implementation can begin. This step may be a "simple" matter of buying some hardware, writing some programs and integrating them into the process. Or the system may be large and complex and may be provided by an outside vendor or contractor on a turnkey basis. In any event, certain guidelines should be followed during the implementation phase.

First, the procedures to be followed must be established, particularly on large projects. Procedures include the definition of communication channels, standards and guidelines, quality control procedures and schedule, to name a few. Again, it is imperative that these be written down to ensure understanding by all parties involved.

Second is the final definition phase. If the system is being supplied by a vendor on a turnkey basis, including hardware and software, this takes the form of submittals, or shop drawings. If the manager has purchased hardware on a purchase order and is doing internal development of the software, this takes the form of detailed design specifications. Whatever the source, this checkpoint is the last one before the details in the specification are "cast in concrete." A careful review of the implementation of the written design, before it is programmed or built, can save a great deal of time later on. One should not wait until it is in the plant to find out that something was overlooked or misunderstood.

This is particularly vital with respect to software because of the inexact nature of its definition. Even if the programming is being done in-house, the manager must insist on a complete description of how the software requirements are being interpreted and met. Contrary to popular belief, software changes in the field are expensive.

A third aspect of installation is checkout and testing. An integrated test plan is essential. The more complex a system, the more critical and time-consuming this step becomes. However, nothing is more valuable to the project than catching problems early so that their impact can be lessened.

A fourth phase is training. At a minimum, operators and maintenance personnel must be fully trained in the use of the system; however, if the system is at all complicated, someone must be aware of the software included and the necessary steps to troubleshoot suspected software bugs. One should not overlook management, which should be aware of the system, its capabilities and its limitations. In short, training must be addressed, and it may be more than a simple matter of showing someone what buttons to push.

Last, but not least, is the system documentation, probably the most overlooked aspect of implementation. In the long run, the system will be only as good as the paperwork behind it. One must not short-change this effort, even if it means going over budget at the end. It will pay dividends many times over.

Success Step 4 — Staffing (Chapter 10)

This chapter deals with the necessary staff for a control system and approaches it from a large system point of view. Large systems usually require more staff to support them.

The chapter is built around the concept of the system manager, the key ingredient in any computer-based control system. Without this person, a system has little chance for success. It may be a full-time position or a part-time assignment for a staff member, depending on the system size and complexity. However, someone must be given the responsibility to manage the system, as well as adequate time and authority to perform this function.

Other considerations, such as staff positions, organizations and contingencies, are dealt with also. The successful manager must spend a good deal of time on this phase of the project.

Success Step 5 — Management and Optimization (Chapter 11)

Once the system has been planned, designed, implemented and staffed, it is only natural to assume that the project is complete and involvement is ended. In fact, the budget probably is expended and the people who have worked so long and hard undoubtedly are being assigned to new endeavors. The system is in and operating. Everyone is reasonably happy with the results and the manager is to be complimented.

However, this is merely the beginning of the final step of the project, one that will be ongoing. It deals with the continuing management of the system, as well as with the aspect of control system optimization. This step never really ends. The prudent manager must recognize that the system is not static and, as much as any dynamic entity, must be managed and controlled.

First, the manager must develop a structure that deals with the entire control system. The manager who places too tight a scope on the problem by dealing only with the computer or the hardware or the software fails to recognize the importance of all of the components to ultimate system success. All the pieces and technical expertise required by them should come under the manager's sphere of influence.

The manager must not leave the procedures for management of the control system to chance. Care should be taken to clearly define and document every procedure, responsibility and authority. These procedures, like every other phase of the project, should be written and distributed. Misunderstandings will be avoided, including overlapping of duties and duties no one claimed. More importantly, staff turnover will have less impact because a well-documented procedures manual will clearly define the incumbent's duties.

Related to the ongoing management of the system is the aspect of process optimization. This deals with the need to "fine tune" the system over time and includes such tasks as loop tuning, control program changes, modifications to operator displays and perhaps alterations to the control philosophy itself. These efforts must be planned and coordinated by the management system to improve the functionality and benefit.

CONTROL SYSTEM PHILOSOPHY

Before embarking on a project that will involve computer process control, someone must address a basic philosophical question: "How much computer control will be included in the system?" This decision must be considered carefully because it will determine the direction of the project from the very early stages of Step 1. A change in this decision later on either may force the project back to the very beginning or, at a minimum, will compromise the overall integrity of the system design.

Although the spectrum of the computer involvement contains a nearly infinite number of possibilities, there are basically four major types of systems, or system philosophies, that can be considered. These are:

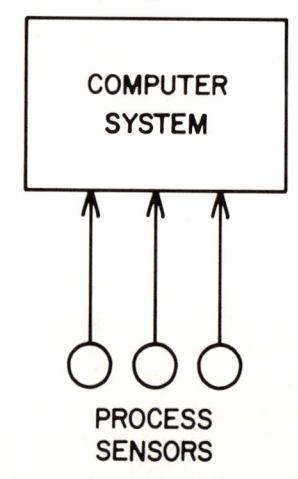

1. Data Acquisition Systems (DAS)
2. Supervisory Control and Data Acquisition (SCADA)
3. Digitally Emulated Analog Control (DEAC)
4. Optimizing Digital Control (ODC)

Data Acquisition Systems (DAS)

The most basic type of computer application is the data acquisition system, as seen in Figure 1. As its name implies, the DAS is used strictly for bringing process information into one location for use by a human operator. The information is stored in computer memory or on a disk. No direct control actions are made by the computer, nor can the operator make any control outputs through the computer. All changes to the process are made manually by an operator or by local analog control equipment.

The information can be utilized in a variety of ways. It can be displayed on a VDT, thus giving the operator a continual window to the process. It also can be logged on a printing device, either continually or when exceptions occur. Abnormal conditions can be annunciated. Of course the data can be processed and summarized in printed form in operating logs and reports.

The basic function of the DAS approach is to centralize all information

Figure 1. Data acquisition system (DAS).

to make easier, faster and better control decisions. However, the computer is not involved directly in the control actions.

Supervisory Control And Data Acquisition (SCADA)

Figure 2 represents a typical SCADA system. The typical functions of a data acquisition system can be found in this approach. The computer still inputs field data for alarming and reporting and for operator information. However, this approach also includes outputs from the computer to the field, as well as the inputs of sensed information.

In a SCADA system, the computer outputs control signals only in response to an operator directive or command. No automatic control actions are taken by the computer. These operator commands usually are initiated through some input device, such as a VDT. However, some systems utilize local panel pushbuttons or thumbwheel inputs. This is the key element of the SCADA system; all outputs occur only as a result of operator commands.

The outputs usually are directed to one of two locations. One is an end element, such as a motor contact or valve actuator. This is common for

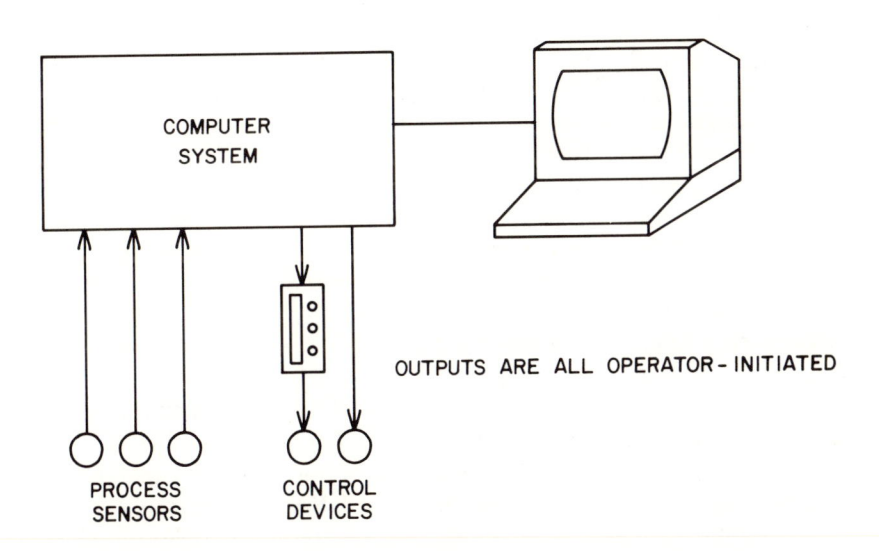

Figure 2. Supervisory control and data acquisition (SCADA).

both discrete or digital signals (turn on or off) and for continuous or analog signals (position). The operator commands directly affect a control device.

The second output destination commonly used is the traditional analog controller. In this case, the outputs from the computer are desired process values or setpoints. The automatic control is accomplished by the analog device. The operator selects a new setpoint value, the computer outputs the value to the controller, and the controller adjusts the end element to maintain the process value. Control is independent of the computer.

Digitally Emulated Analog Control (DEAC)

Digital systems that are designed to perform automatic process control without operator intervention and without the aid of analog hardware (controllers) usually are termed Direct Digital Control Systems (DDC). However, the world of DDC generally can be separated into two varieties. The first and most basic is the DEAC approach (Figure 3).

The DEAC approach to DDC merely replaces the control usually performed in analog hardware and relay logic with programs written in a digital computer. This means that conventional control loops are included, made up of three-mode proportional, integral and derivative (PID) controllers, timers, lead-lag, deadtime compensators and sequencers (relay replacement), among others. The modern programmable controller fits this category, as do most DDC systems installed today.

The advantages of this approach are simplicity, cost reduction and improved reliability. The hardware cost is frequently less than analog, and the data acquisition capabilities provide added benefits. However, the resulting control is generally no better and no more efficient than that available through an analog approach.

Optimizing Digital Control (ODC)

The second type of DDC system is the ODC type approach (Figure 4). It is one step in sophistication beyond the DEAC form of DDC. This type of design strives to take advantage of the superior capability of digital logic and software by building on a DEAC foundation.

In an ODC-designed system, the basic regulatory modules of DEAC have a higher level of logic, which performs some form of optimizing function. This may include a control strategy for optimizing resources

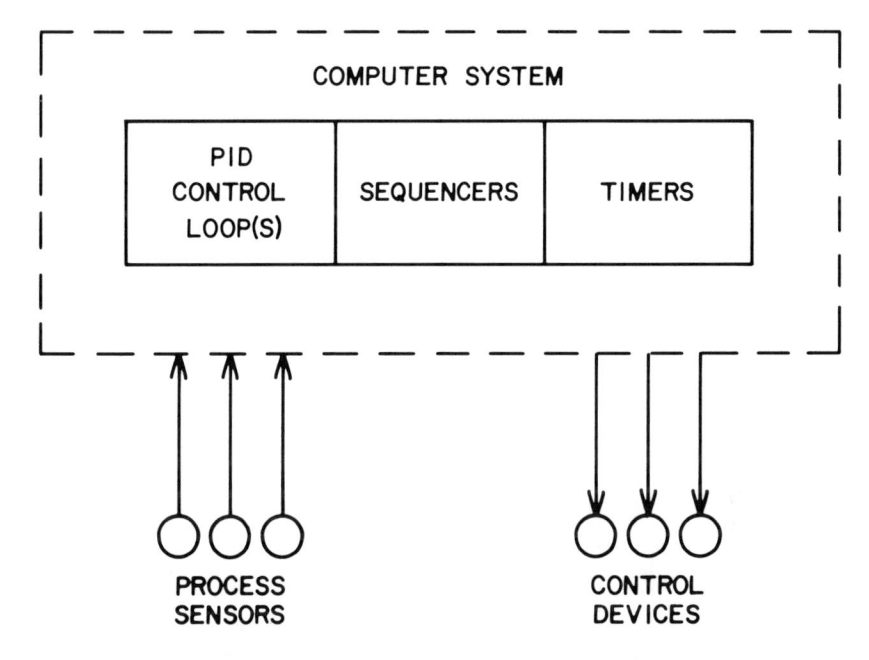

```
┌─ ─ ─ ─ ─ ─ ─ ─ ─ ─ ─ ─ ─ ─ ─ ─ ─ ─ ─ ─ ┐
│            COMPUTER SYSTEM             │
│  ┌──────────┬──────────┬──────────┐    │
│  │   PID    │          │          │    │
│  │ CONTROL  │SEQUENCERS│  TIMERS  │    │
│  │ LOOP(S)  │          │          │    │
│  └──────────┴──────────┴──────────┘    │
└─ ─ ─ ─ ─ ─ ─ ─ ─ ─ ─ ─ ─ ─ ─ ─ ─ ─ ─ ─ ┘
```

Figure 3. Digitally emulated analog control (DEAC).

for a process or may be a combination of programs coordinating actions in various plant areas. One significant potential for ODC is energy management, including power factor correction, load deferral or even load shedding when demand charges are to be avoided.

The ODC design concepts offer the greatest opportunity for benefits relative to digital control. However, these major savings are not achieved without a cost and are not readily attainable unless the user properly pursues the project, including planning, design, implementation, staffing, management and optimization. With proper attention to detail and with adequate management commitment and direction, significant savings can be realized.

TECHNICAL HELP – THE MANAGER'S OPTIONS

Basic to the success steps identified by this book is the implied premise that the manager is technically competent or has staff who can address the many and varied aspects of computer control. Success step 4 (Chapter 10) deals with staffing; however, that chapter's emphasis is staffing

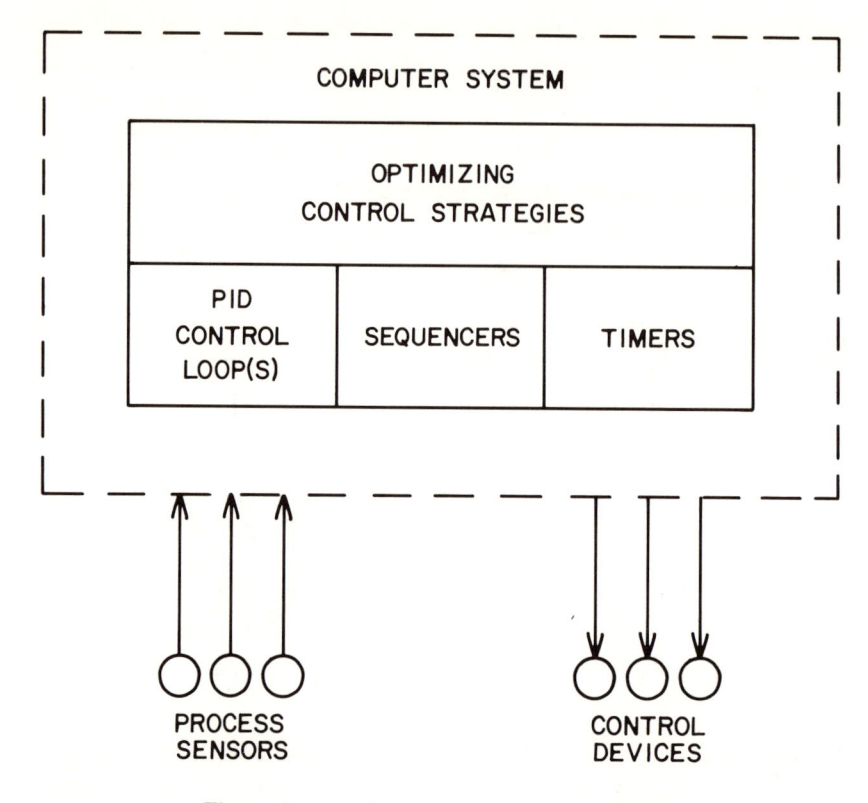

```
┌──────────────────────────────────────────┐
│             COMPUTER SYSTEM              │
│  ┌────────────────────────────────────┐  │
│  │          OPTIMIZING                │  │
│  │       CONTROL STRATEGIES           │  │
│  ├──────────┬──────────┬──────────────┤  │
│  │   PID    │          │              │  │
│  │ CONTROL  │SEQUENCERS│    TIMERS    │  │
│  │ LOOP(S)  │          │              │  │
│  └──────────┴──────────┴──────────────┘  │
└──────────────────────────────────────────┘
```

Figure 4. Optimizing digital control (ODC).

required to support the system once it is functional. What does the manager do when faced with the prospects of a digital system project without the staff necessary to do steps 1–3? Generally, there are two options.

Hiring Additional Staff In-House

One very definite possibility in many instances is internal hiring. The addition of technical staff has a certain desirability. It may provide better control than using "outsiders," it can protect any type of proprietary processes to be controlled or monitored; and it can permit in-house process expertise to be applied with a minimal learning curve.

But where does one start to build an in-house capability? First, the manager must decide how big a staff is needed and what capabilities are required. This is probably the hardest part because the complexities will not be readily apparent unless one analyzes the situation very carefully. Chapter 10 is a good starting point because the staff needed to support the system will, in many instances, be helpful in planning, design and implementation. However, there probably will be special requirements for these early phases which the continuing staff will not necessarily have to provide. Let us take each step separately.

Planning

The basic need in the planning stages is a clear understanding of the problems and objectives to be attained and the various technical solutions to be considered. Generally, the needs and objectives definition requires somebody who can listen and ask questions of the appropriate people, drawing the correct conclusions and synthesizing the common requirements. This person must have a basic understanding of the process, as well as a good background in digital technology and instrumentation systems. However, good communication skills are far more important than technical expertise because process people undoubtedly can address these details. Once the problems and objectives are clear, the real technical expertise is required. This will vary by project but generally will require someone with knowledge of the computer hardware and software offered by various suppliers who manufacture systems that may meet the needs. The manager also needs expertise in interfacing the computer to the process including sensors, instruments and control devices. Depending on the extent of the project, one also must be concerned with control room planning, data communications and telemetry, operator-process interface and, of course, process control theory.

One word of caution is in order at this point. It is the rare individual indeed who will satisfy all these needs. One must be prepared to hire a team of people, particularly if one is planning to go on to the design phase. Although one person may be found who knows enough to perform the basic planning functions, it is doubtful that he/she will know enough to perform the design in the detail required. Furthermore, any project of any size will require more than one person just to finish in any reasonable time.

Design

All the capabilities given above must be satisfied by the project team. At this point, however, the manager requires a very detailed capability, much more so than in the planning phase. The people must be able to translate general requirements into detailed drawings and specifications.

This involves all aspects of the project, from process to computer. If the system is being applied to an existing process, the details of field installation without process disruption must be worked out. Careful evaluation of new instrumentation is necessary. If the system is to be bid competitively, the "specmanship" is very important. Above all, the detailed definition of the software must not be overlooked.

Again, team capabilities must be considered separate from team size. The manager needs the technical expertise and also enough people to meet the project timetable. Detailed design, when properly done, can take considerable time unless enough people with the correct combination of talents are involved.

Implementation

Staff requirements for implementation will depend largely on how much of the system the supplier will implement (versus the staff) and how complex is the design. Obviously, if the vendor does a great deal, the manager's needs are fewer. Similarly, the small programmable controller requires less staff to implement than does the large, processwide control system.

If the system is to be implemented by the manager and is of a moderate complexity or more, one should be considering a fair-sized staff. It is not unusual to need many man-years of effort to implement a system. If one is buying hardware and writing the software from scratch, this effort can stretch out considerably. If one can make use of various vendor-supplied "packages," less involvement can be expected. Nevertheless, the manager's commitment needs to be considerable unless one has a long time to get the project operational.

If a vendor is supplying a turnkey system, staff involvement will be less, at least in terms of staff size, because the vendor provides the needed manpower. However, the manager must have the same technical expertise to verify the adequacy of the vendor's system, both during the implementation and at the final system test. In short, one needs the same capabilities, just fewer of them!

Obtaining Outside Assistance

Instrumentation and controls (I&C) generally is considered a specialty field. Many companies do not want to hire a large staff of specialists for one project and then either "carry them" until the next project comes along or lay them off. Other companies have fairly frequent projects but do not want to pay the cost of keeping a staff up-to-date in the state-of-the-art, particularly when the technology involved is so frantically changing.

However, "going outside" is not an easy proposition. Care must be taken to select a consultant who can meet all the needs discussed above and in a timely fashion. Although the selection process can take several forms, the following qualifications are essential.

Relevant Experience

The design professional must understand the technical aspects of the project. The manager must first understand the needs and then look for direct experience in all areas. One should read their literature and check their references. Ask about directly related projects and then check them out. The manager should not be influenced by a general brochure or smooth presentation without checking on the details.

Staff Availability

One should be sure that the firm has the right people and that enough are available to work on the project. The manager should not hire a one-person firm who promises to bring on staff, unless the project only needs one person. It will take time to find those new people, and you will pay for their learning curve.

Vendor Independence

Vendors know and understand their equipment. They may even know a fair amount about the process in question. However, no equipment is the best for *all* applications. Therefore, the design professional should be free from any ties to an equipment supplier and should be someone who can *objectively* analyze the needs and then design the best solution to the problem. If the designer is related to a vendor, in any way, it is certain that the "best" solution somehow will involve that vendor's equipment.

SUMMARY

The manager who is charged with overall responsibility of a computer-based control system faces a complex task if the expectations of all concerned are to be realized. However, this book offers a series of steps that, if carefully carried out, will make this task less monumental.

Before beginning, the manager requires some basic familiarity with the technology. This includes the differences between process control and data processing, the basics of hardware and software and the reasonable benefits to be expected from the system. An appreciation of the people aspect is also necessary.

The five formal steps presented by this book will then form a structure for project success. It must be planned, designed, implemented, staffed and managed. Each step must be completed and documented. No shortcuts can be taken without considerable risk.

The manager must realistically evaluate the staffing required to complete the five steps. If adequate capabilities are not on staff, the prudent manager must evaluate hiring additional personnel versus retaining a consultant for the project. Care must be taken in selecting new personnel, as well as in selecting a consultant.

REFERENCE

1. Harrison, T. J. *Minicomputers in Industrial Control* (Pittsburgh, PA: Instrument Society of America, 1978).

CHAPTER 2

PROCESS CONTROL IS NOT DATA PROCESSING

Dag I. Knudsen, PE

INTRODUCTION

"Computerize It" — A Modern Word for Problem Solving

Horror stories abound about problems blamed on the computer. Yet, we know the computer can be an effective tool. If the computer is indeed an effective adjunct to modern process control, why are there so many less-than-satisfactory instances of its application?

Bad experiences with computers in process monitoring and control applications are most often traced to people. It seems that highly qualified and articulate computer experts are "selling" computers into process control applications. Neither they nor their customers realize that the user lacks the practical knowledge and realism about the process control environment, a knowledge essential to the success of the application. It is characteristically human to tackle a task on the basis of immediate past experience. However, data processing experience with financial and management information system (MIS) applications is inadequate for process monitoring and control applications.

The purpose of this chapter is to present the characteristic differences between data processing and process control (PC). The discussions will center on differences in application, equipment, software and people. Although the basic computer system hardware elements are similar for data processing and process control, there are differences. The same is true for software. But in the end, people represent the real difference.

"Cobbler, stay with your last" because what you do well today is based on your education and immediate past experience. A data processing analyst/programmer must understand the application very well to design a logic system that satisfies the needs (payroll, accounts receivable, etc.). A control system engineer must also understand the applications. These applications are twofold. Understanding the process to be controlled is important to ensure safety of equipment and personnel and to know the relationships between measured and manipulated variables. Second, understanding the available tools is necessary to achieve the results that are possible.

Cost reductions and profit improvements can be achieved with computer-based automatic control systems, provided that the following requirements are met:

1. Technical expertise with *relevant* experience should be involved in the planning, design, procurement, implementation, startup and operation.
2. Equipment suited for the application should be procured based on its ability to meet needs that must be defined in detail prior to selecting a vendor and equipment.

APPLICATION DIFFERENCES

The Corporate Model

There are numerous applications for computers in an organization. The corporate model (Figure 1) shows a typical organization consisting of three elements:

1. Mainstream activities involve producing a product or result that represents the organization's "raison d'être."
2. Administrative activities provide leadership and financial record-keeping.
3. Supportive activities involve marketing, planning, engineering and research.

Table I classifies these activities into two groups. One group utilizes electronic data processing as a tool. The second represents applications for machine automation and closed-loop automatic process control.

Data Processing Applications

Early computer applications were found in the administrative and supportive business areas, even though the cost-saving potentials were not as

```
                        MAINSTREAM
                            │
                            ▼
                  ┌─────────────────────────┐
ADMINISTRATIVE    │  SALES ORDER PROCESSING │    SUPPORTIVE
     │            │  INVENTORY CONTROL      │        │
     ▼            │  REQUIREMENTS GENERATION│        ▼
┌──────────┐      │  PRODUCTION SCHEDULING  │   ┌──────────────┐
│ FINANCIAL│      │  PRODUCTION CONTROL     │   │  MARKETING   │
│ PERSONNEL│ ◄──► ├─────────────────────────┤◄─►│  PLANNING    │
│ LEGAL    │      │  MACHINE CONTROL        │   │  ENGINEERING │
│          │      │  PROCESS CONTROL        │   │  RESEARCH    │
└──────────┘      ├─────────────────────────┤   │DATA PROCESSING│
                  │  PURCHASING             │   └──────────────┘
                  │  RECEIVING              │
                  │  QUALITY CONTROL        │
                  │  MAINTENANCE            │
                  └─────────────────────────┘
                            │
                            ▼
```

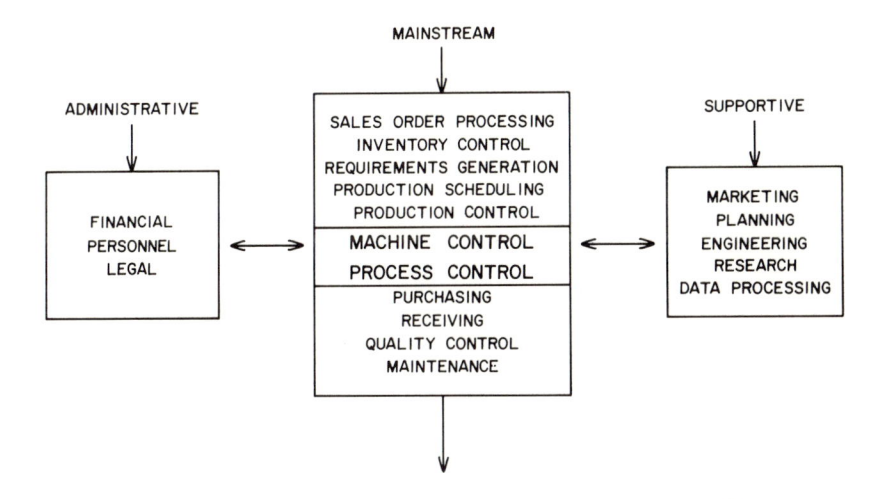

Figure 1. The corporate model.

high as elsewhere (Table II). A 10% cost reduction has the greatest impact when achieved in the mainstream activities [1].

These applications represented progressive improvements in the automation of office functions such as payroll, accounts receivable (AR), accounts payable (AP), general ledger (GL), cost accounting, profit and loss statements (P/L) and budgeting. As organizations grew in size, their ability to produce payroll and be up-to-date on their financial status on a timely basis could not be met by merely adding people. Thus, the need for automation was there.

The need for better planning, scheduling and information availability became increasingly important in the mainstream activities of most companies. The electronic digital computer had proven itself to be a reliable and valuable part of the administrative and support areas, and those benefits could be utilized to improve productivity, inventory availability, quality and efficiency. The evolution was quite natural and includes such applications as inventory control, bill of materials (BOM), material requirement planning (MRP), production scheduling, automatic order entry and tracking, route optimization and carrier selection, maintenance planning, scheduling and control, and cost control planning and scheduling, to name a few.

In general, electronic data processing (EDP) involves applications where people are the source of inputs (e.g., items in/out of inventory). The outputs are in the form of schedules, job orders and reports. These too are handled by people. This means that people are involved in both the data input and data output sides of the system. The modern interface

22   PROCESS CONTROL COMPUTER SYSTEMS

Table I.   Computer Applications for Automation

	EDP	PC
Administrative		
Financial (AR, AP, GL, cost accounting, payroll, etc.)	X	
Budgeting	X	
Financial models	X	
P/L	X	
Supporting		
DP department		
Sales analysis, forecasting, etc.	X	
Planning (economics, decision, operations research)	X	
Computer-aided design (CAD)	X	
Project control and Scheduling	X	
Mainstream		
Order processing	X	
Inventory control (MRP)	X	
Production/machine control (numeric control)		X
Process control		X
Maintenance	X	
Manufacturing	X	

Table II.   Cost Distribution for a Typical Processing/Manufacturing Company

Administrative Costs	5
Supportive Costs	15
Mainstream Costs	80

consists of a video display terminal (VDT), keyboard and a printer.

A second key element of EDP is the data base. It is only an approximation of the current state of the physical system it represents. For example, consider a system that plans and schedules routine equipment maintenance. The computer schedules a particular task for a specific day and prints a work order. Its data base reflects the fact that the maintenance is scheduled but has not been done yet. It will continue to show the work as uncompleted until a person updates the data base with the completed work order information. In short, the data base for EDP applications is generally not "real-time" because it does not reflect the true status of the physical system. This is not to say that it is not accurate or worthwhile; rather, it merely points out a key difference between EDP and process control.

Process Control Applications

Automation may be an automated payroll system, an automated machine, an automated assembly line process, an automated startup/shutdown sequence or automatic process control. Each form is different, requiring different know-how and experience, even though a computer may be used in each instance. The various types of automation are distinctly different in terms of the computing functions.

Automated payroll depends on people-based inputs and outputs. The computer function is that of automatic processing of data. Data inputs add to an existing data base. Data files, representing historic records, expand with time. The computer functions as a calculator and data storage/retrieval device. It is the functional equivalent of a file cabinet.

Automation also can be based on automatic sensing of process variables (status, position, speed, temperature, pressure, etc.), automatic calculations of control actions to be performed and automatic control outputs to the control devices (manipulated variables). This is what is meant by the term automatic control and includes such applications as:

1. robotics (on a production line);
2. numerically controlled machines (drill presses, metal stamping, shaping and forming machines);
3. logic control for repetitive operations (e.g., position, fill, cap on a bottling line);
4. sequence control for automatic startup and shutdown of plants (e.g., electric generating unit) or process units (e.g., an incinerator or steam generator);
5. batch control for blending and mixing operations; and
6. process control, used to maintain specific process conditions (e.g., flowrate, temperature, acidity, etc.).

The forms of automation listed above represent a type distinctly different from automated data processing. They replace people functions, may or may not depend on people-based inputs and outputs, and have a fixed, nonexpanding data base. There are also major differences within these forms of automation. Two basic distinctions are discrete control and continuous control.

Discrete Control

Discrete control correlates equipment status (on/off) and status changes with a predefined program of events and time. Discrete logic control may be represented by a sequence of operations performed in a manufacturing process, e.g., bottling, canning, clean-in-place (CIP) systems. Discrete sequence control also may be represented by an automatic startup/shutdown operation. The operation may be initiated manually by an operator (e.g., push start button) or automatically by an external event (e.g., burner flame-out results in an orderly boiler shutdown).

Programmable Logic Controllers (PLC) have become the hallmark of modern discrete control capability, although discrete logic and sequence control also can be performed in minicomputers. The PLC emulates relay ladder control methods. The need to wire and rewire combinations of relays and timers is eliminated by the nature of the PLC's programmable characteristics. Being programmable means that the PLC can accept data to alter the state of its internal circuitry to perform different specific tasks.

To program a PLC for a specific application requires experience with the design of relay ladder networks because the programming tools emulate the hardwired method directly. Logic algorithms are programmed into minicomputers through the use of Boolean algebra techniques. This method requires more experience with logic design, but also offers more capability.

Continuous Control

Continuous control may be recognized by many names: setpoint control, regulatory control, analog control, feedback control, process control, automatic control or industrial control. More technical terms are often used such as P, PI or PID control. These designations reflect specific features of the continuous type of controller. P refers to proportional control action, PI refers to proportional and integral control actions, PID refers to proportional, integral and derivative control actions. P, I and D are the basic tuning constants that need adjustment for the controller to maintain the desired setpoint. Process and measurement deadtimes and response times determine the type required. Figure 2 shows a typical analog controller and the symbology used to show it on control diagrams.

Continuous control, in its basic form, is called continuous because it depends on a measurement for its input and manipulates a control device

PROCESS CONTROL IS NOT DATA PROCESSING

as its output. In Figure 3, the process variable (flow) is sensed by measuring the differential pressure drop across a venturi tube. The differential pressure transmitter converts the pressure drop into an equivalent (analog) electrical or pneumatic signal for transmission to the analog controller. The analog controller compares the process variable (PV) to its desired rate of flow (i.e., setpoint) and calculates the change required (absolute or incremental depending on the type of controller), if any, and changes its output accordingly. The output may be electrical, pneumatic or hydraulic.

Hybrid Control

Hybrid control combines discrete and continuous control into control system strategies. These strategies enable security, safety and optimum

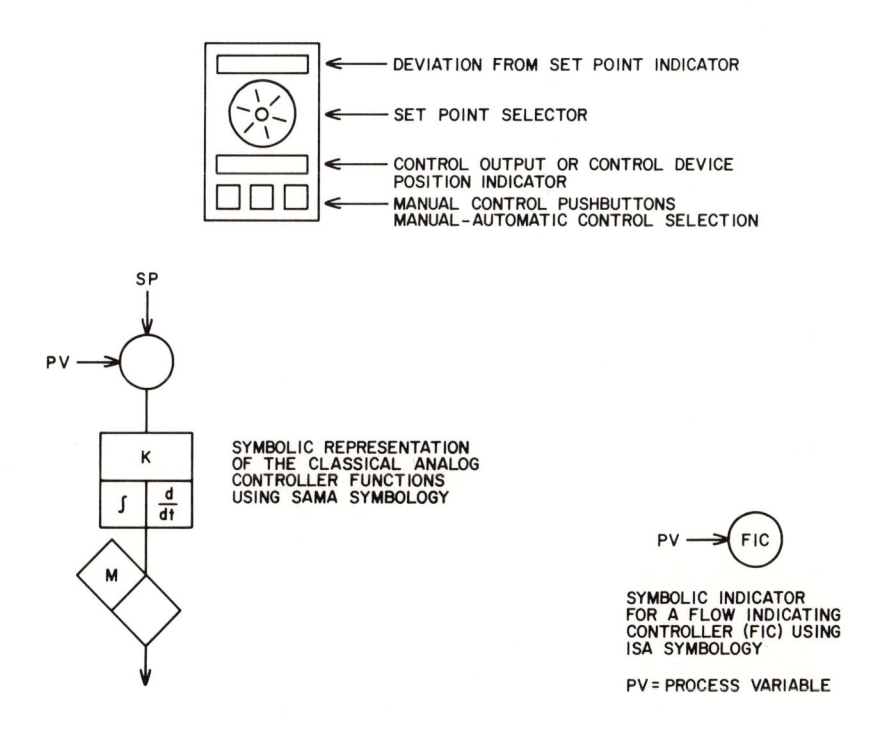

Figure 2. The classical analog controller.

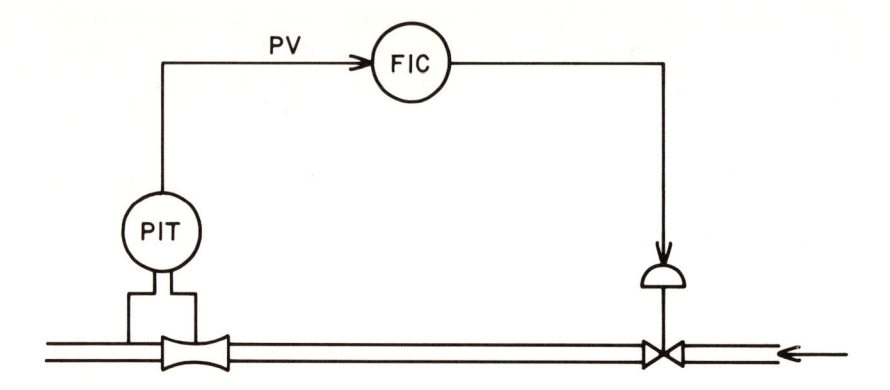

Figure 3. Flow control loop.

performance. The computer can be used effectively for hybrid control strategies. Startup and shutdown logic, failure detection (instruments and equipment) and process characteristic tracking can be used to minimize dependence on operator-directed corrective actions and maintain performance.

Summary

In summary, process control applications generally operate in "real-time," based on sensor inputs and physical outputs. People are not necessary to provide inputs or to execute outputs. The data base is fixed and always represents exactly the physical system. While EDP applications are generally involved in collecting, storing and manipulating a great deal of data for human use and application, process control applications usually involve a relatively small amount of data, with physical actions by the computer as the ultimate goal.

EQUIPMENT DIFFERENCES

Figure 4 outlines the basic computer hardware elements for a data processing system. All data processing systems include combinations of these basic elements. Figure 5 outlines the hardware components in a computer-based process control system. Data loggers/monitoring-only systems do not include the analog and digital output multiplexers. Otherwise, all systems include these basic elements.

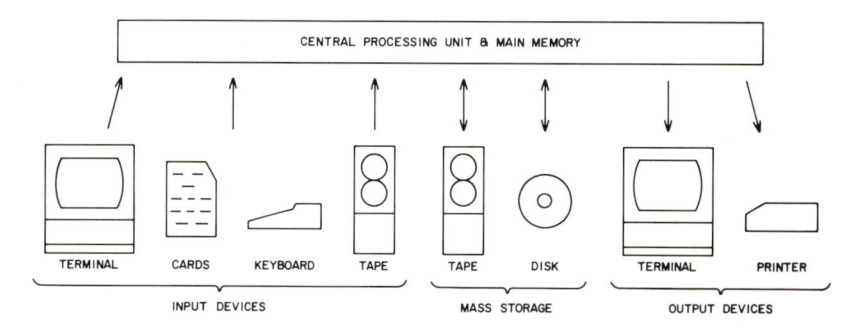

Figure 4. Computer hardware components for a data processing system.

The Central Processing Unit

The Central Processing Unit (CPU) is the central element of all computer systems. A CPU's basic capability is defined by four characteristics. These are cycle time, word length, instruction set and architecture. Cycle time is the minimum time required to read and execute a single instruction. Word length can be 4, 8, 16, 24, 32, etc., bits long. Instruction set is the set of functions that can be performed by the CPU itself. Architecture is defined by the number of registers, number of processors and bus structure.

The CPU's characteristics determine its application. A four-function calculator, for example, is represented by a CPU that can perform the functions of addition, subtraction, division and multiplication. Additional instructions might be added, such as square root and percentage. Adding a storage register results in memory capability. An operator can perform complex calculations with a simple calculator if the techniques are known.

The CPU used in process monitoring and control applications usually is not the same CPU used for a data processing application. A CPU, like any other physical device, may be designed to perform well in a specific application. The CPU in a PLC may have capabilities different from that used in a process controller.

Special-purpose CPUs result in lower cost, but also result in reduced capabilities. A buyer's expectations should be moderated accordingly. Because the modern distributed controllers have capabilities more limited than earlier minicomputers, it is more important than ever to design the application before the hardware procurement is made. Several suppliers of "computer-based control systems" have CPUs that are specifically manufactured for their product.

28 PROCESS CONTROL COMPUTER SYSTEMS

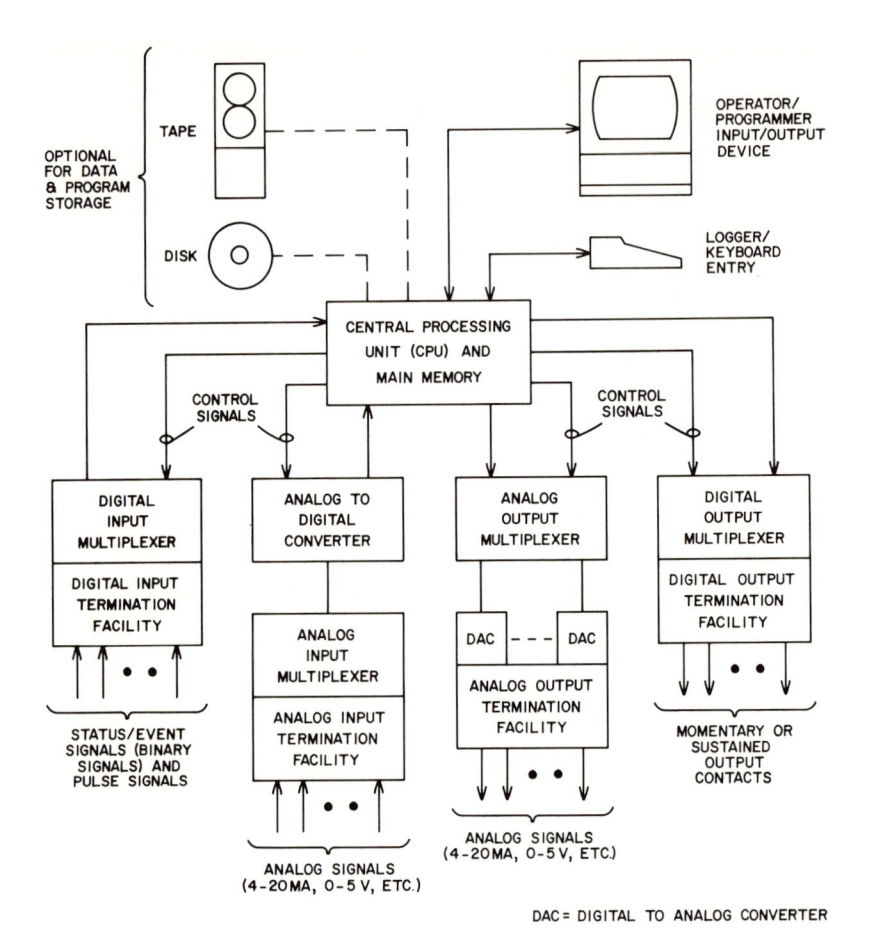

Figure 5. Computer hardware components for process monitoring and control systems.

Computer Memory

Typical volatile memories include solid state. Nonvolatile memories are magnetic core and bubble type. Nonvolatile memory is used where loss of data and instructions is undesirable during power outages. Solid-state memories are made nonvolatile by the use of special battery backup. Either type can be used in the typical process control system. However, their selection is based on the specific needs of the installation. These needs may be reliability, the likelihood of potentially damaging electrical transients, high-temperature environments or vibration.

High-speed memory devices are expensive. Data and programs are therefore stored on mass storage devices where the cost per character is less by several orders of magnitude.

Online storage is achieved with magnetically coated discs. Offline storage is achieved with removable discs and magnetic tape units. Capacity, access time, reliability and cost vary greatly. Process control systems may be configured without mass storage devices. This enhances system reliability. Where mass storage is required, they are selected to be highly reliable. This may include units purged with inert gas and/or nonmoving sensing heads. Redundant configurations also are used to increase overall system reliability. Process control management systems have expandable memory needs.

Data processing (DP) applications always require some form of auxiliary memory. Furthermore, memory requirements, both main memory and auxiliary memory, are typically larger. While process control systems often need only 32,000 to 64,000 bytes of memory, it is not unusual to find even small DP systems with 256,000 bytes or more. Auxiliary memory differences are even more pronounced.

Input/Output (I/O) Devices

Programs and data are entered into the computer through an input device. Batch-type data processing systems normally utilize card readers, and/or key to tape/tape recorders. Online DP systems utilize terminals consisting of a VDT screen and keyboard. Information stored in the computer also can be accessed and displayed on the VDT screen.

Data entry to a process control computer is an automatic function. The computer scans the digital and analog input systems periodically (e.g., every 1, 10 seconds, etc.) and updates respective memory locations with new data. The old data are discarded (except in management-type control systems). The operator input device enables an operator to access the system to determine its status and to change operational parameters as required. This terminal does not serve the function of input of data per se, as with a terminal in a DP system. Printers are used where hardcopy records are required. VDT terminals are used where an inquiry without hardcopy is desired. These may be common to both.

SOFTWARE DIFFERENCES

To be useful, the computer must be instructed in what to do and how to do it. These instructions are contained in programs, or software. A program is the set of instructions needed to perform a specific task. The instructions are executed in the order they are written. Any application requires a collection of programs, and this collection forms a software system.

Operating Systems

An operating system (OS) is the nucleus of the computer software system. It controls, monitors, executes, schedules and loads programs, produces the computer log, controls multiprogramming (the apparent simultaneous execution of two or more programs) and handles routing and scheduling of terminal communications. A given computer may be available with more than one OS. The intended computer use determines which OS is applicable. The OS is usually developed and supplied by the computer manufacturer; however, computer system users sometimes develop their own OS to meet specific requirements.

Programs, once started, run to completion in batch-type DP operating systems. A time-slicing technique is used in multiprogrammed EDP systems. Each active program runs for a fixed time period (e.g., 50 milliseconds).

A PC system literally has its "hand on the throttle" and must respond accordingly. The real-time process control operating system must respond to external and internal events in a timely fashion. An external (hardware) event, or interrupt, may be activated by a digital input that changes state. Its priority, relative to all interrupts, determines the action required. If the priority is higher than that of the program in progress, it causes the program to suspend operation and the relevant program(s) to commence. The suspended program resumes operation when higher-priority activities are completed. This is all controlled and managed by the OS. This is quite different from the DP requirement, although some DP applications have been implemented on PC operating systems; however, the converse is seldom, or never, the case.

Programming Languages

The application software is a collection of programs that does a useful task for the user. This may be payroll, accounts receivable, process control or scientific applications such as mathematical modeling.

A programming language is a set of rules and conventions used to prepare the source programs for translation by the computer into machine executable form. Programming languages generally are divided into

PROCESS CONTROL IS NOT DATA PROCESSING 31

Table III. Higher-Level Programming Languages

Year Introduced	Name	"Label"	Application
1956	Formula translator	FORTRAN	Scientific
1956	List processor	LISP	Artificial Intelligence
1958	Algorithm language	ALGOL	Scientific
1960	Common business-oriented language	COBOL	Business
1964	Programming language 1	PL/1	Business, scientific
1965	Beginners all purpose Symbolic instruction code	BASIC	Miscellaneous
Mid-1960s	Process control language(s)	–	Process control
1971	Blaise Pascal	PASCAL	Scientific, process control
Mid-1970s	Ladder diagram language(s)	–	Relay replacement control
Early 1980s	Standardized real-time control language	ADA	Real-time DP and automation

high- and low-level languages. Assembler is a low-level language. It requires greater skills on the part of the programmer and takes longer to prepare, but usually requires less memory and executes faster. High-level languages have fewer rules and result in faster program production. Several are available. A given computer may accommodate one or more of the high-level languages.

The high-level languages differ in the type of data they can handle, operations they can perform on data and control functions they provide for structuring programs. More than 150 programming languages are in use in the United States. Some of the more commonly used languages are listed in Table III [2].

Computers generally are considered to deal with numbers (integers) and their manipulation (calculations). This is true for scientific and control applications; however, data processing applications predominantly use data elements of a different kind. For example, in the filing system for a company payroll record, the individual employee's record consists of the employee's name stored as a character string, his/her employee number stored as an integer data element and payment records stored as floating-point numbers. The computer files of a business organization involve thousands of different kinds of information/data, each having a distinct type and format. Business applications place less emphasis on complex numerical applications and more emphasis on techniques for storing, arranging and retrieving large quantities of data. It is only natural, then, that the programming language's emphasis is on storing, arranging and retrieving data and, hence, the use of COBOL and BASIC in business-oriented electronic data processing.

COBOL is the most widely used language for commercial data processing. It can handle large amounts of data and perform the calculations required for such tasks as financial recordkeeping. COBOL includes extensive input facilities to structure information and output facilities for report generation. Being intended for everyday data processing, it was designed to be closer to English than other programming languages.

Data Base Management Systems (DBMS) represent a more recent programming tool. Containing a large collection of data processing programs, DBMS is based on fixed ways of structuring large quantities of data within a computer system.

Other languages are developed specifically to accommodate the needs of engineering and scientific applications. These also free the programmer from concern with binary strings, memory addresses and basic computer operations. These programming languages enable engineering programmers to concentrate on the computing problem to be solved. FORTRAN (formula translator) is the most popular algebraic language designed specifically for scientific applications.

Specific languages also have been developed for process control. Some are derivatives of FORTRAN, others are oriented to special problems or applications. The process control languages (PCL) offer a range of capabilities and require programming experience that ranges from none to assembly-level programming.

PLC were designed as replacements for conventional racks of electromechanical relays. The PLC programming languages therefore emulate the documentation method used for conventional relay systems. These "relay ladder programming languages" require no programming knowledge whatsoever. The desired functions are directly selectable by the engineer. A callup or fill-in-the-blanks menu format is used to configure the ladder diagrams.

PLC manufacturers have expanded their capabilities into process control functions as a result of the ever-increasing capabilities of the microprocessors and the increasing acceptance of digital process control. However, this acceptance also requires that the engineers/end-users be conversant in higher-level programming languages and automatic process control technology.

Some higher-level languages are problem oriented. Control programs are developed in a manner similar to the relay ladder diagrams. Control algorithm blocks are linked in software through a question-answer or fill-in-the-blanks process. This method represents a software emulation of equivalent hardware functions. The problem-oriented control languages may be specific for process control or batch control.

Programming a control system includes several functions in addition to the basic control software itself. Scanning, conversion, error checking and storage of process measurements is needed to bring process information into the system. Displaying inputs and controllers on the VDT is required to communicate with an operator. Communication with another computer may be required to collect additional process information or transmit current data for long-term data storage.

Optimizing control may require special equations and control functions. This may include linear and nonlinear programming and algebraic manipulation of equations. These capabilities may or may not be present in the problem-oriented languages. The application therefore may require higher-level general purpose languages such as FORTRAN. This then requires considerable knowledge of computer language programming.

Although several problem-oriented languages are available, no one language meets all the needs of process control and process control management systems.

PERSONNEL DIFFERENCES

Data Processing

DP has become a most important administrative service in the modern corporation, whether private or public. In the beginning, the DP activities concentrated in the areas of finance and became a group within finance. As applications expanded into management information and engineering, DP functions became departments of their own. As a service group, they tended to control all aspects of the computer technology within a corporation. As applications moved into other areas of a corporation, new problems were created. The central DP staff did not have relevant application expertise, and computer equipment and available software did not represent an optimum match to the people who knew the application.

The advent of mini- and microcomputers resulted in truly distributed processing, and dependence on a DP department with a mainframe computer was reduced and/or eliminated. Desk-top computers resulted

34 PROCESS CONTROL COMPUTER SYSTEMS

in distribution to the individual user. The individual organizational unit's automation needs are being satisfied with distributed processing; however, the need for adequate staff is amplified because applications now may be found in many departments. Management style and organization will continue to dictate the effectiveness with which this technology can be utilized.

Typical data processing staff consists of the following:

1. **Manager**—an individual with strong systems and management skills.
2. **System Analyst**—an individual with strong application experience in one or more areas. This could be financial, information systems, order processing, inventory control, manufacturing requirements planning, maintenance management or others. The system analyst is skilled in the definition of problems and the development of algorithms for their solution.
3. **Programmer**—a person who translates the system analyst's design into programs. The programmer is experienced in a programming language (COBOL, RPG or similar).
4. **Computer Operator**—a person who operates the computer, loads and unloads disks and tapes, maintains tape and disk inventory and records.
5. **Key Operator**—an individual who keys information received from programmers or data entry forms into punched cards, magnetic tape or direct computer entry through a VDT.

Process Control

The science of process control is unique and distinct from any other form of automation. The continuous control techniques are dynamic. Their successful application requires an understanding of process and measurement dynamics. On-the-job training with chemical, mechanical and sometimes electrical engineering background represents the basic resource. Advanced degrees are available reaching to Ph.D. in control engineering.

The benefits of dealing with people who have relevant long-term experience in control system engineering cannot be overemphasized. This is especially true when computer technology is applied. The computer offers process control capabilities far beyond the capabilities of the classical analog instrumentation.

A computer-based automatic process control system design team usually consists of the following personnel:

1. **Control System Engineer** (CSE)—a person responsible for the system aspects and project management. He/she interacts with the process, electrical, mechanical and architectural designers. The CSE assures

that overall objectives are met and that the process design is compatible with the process control objectives. The technical background may be engineering or computer sciences. The experience is gained on the job.
2. **Instrumentation Engineer** — a person who must understand the science of measurement and be sensitive to the application needs. Proper application and installation of measurement and control devices is his/her responsibility.
3. **Control Programmer** — a person who must understand the principles of process control and computer programming. His/her technical background is typically science or engineering.
4. **Designer** — a person who is involved with various aspects of control system designs, such as control panel layout, electrical interface design equipment listings and wiring schedules. The designer works under the direction of an engineer and is familiar with control hardware.
5. **Control System Analyst (CSA)** — a problem solver who serves as a consultant to the design team and reviews control applications for controllability. Questionable situations are simulated to determine workable strategies.

Logic control engineering may range from simple startup/shutdown relay ladder networks to complex interlocking or optimizing systems defined in Boolean algebra. Electrical designers learn to develop relay ladder designs on the job. Engineers learn how to design logic systems using mathematical techniques.

VENDOR DIFFERENCES

The computer industry is an estimated $30 billion per year business. Numerous manufacturers, systems houses and original equipment manufacturers (OEM) are involved in the effort of bringing the computer to the users.

Data Processing

More than 70 computer manufacturers produce more than 140 different models of minicomputers. These and other manufacturers also produce a variety of large-scale computers and microcomputers. Although several manufacturers have application software available for sale, their primary interest lies in selling hardware. A computer manufacturer's software products have one purpose: to enhance hardware sales.

Several sales organizations are utilized. The manufacturer's sales staff sells directly to end users, software and manufacturing, or system, OEMs. The latter often will integrate the computer into larger systems. The software OEMs typically develop a line of software to run on specific computers. They offer complete system and software support to their clients.

The software support consists of warranties, user training, installation assistance and software maintenance contracts. The maintenance contract provides the user with enhancements as they are developed by the software OEM. The user is assured participation in improvements as they are developed, so in-house programming staff is not required.

The computer manufacturers will provide service for their equipment on a contract or per diem basis. This discussion holds true for data processing type of software and hardware.

Process Control

Process control computers, systems and software require a different approach. A computer-based process monitoring and control system may be obtained from a systems supplier or integrator, or be integrated by the owner.

The system supplier typically manufactures a large portion of the hardware used in the system. The computer may be manufactured by the system supplier. Alternatively, it may be custom built by another company or represent a standard computer manufacturer's product. The system supplier's value added, therefore, consists of hardware, application programming, standard software and system engineering.

The system integrator typically buys all the hardware from other sources. Their value added consists of application programming, standard software and system engineering.

Application programs are always custom written in PC systems. It is in the buyer's best interest to obtain a source code for all application and utility programs. Enhancements are project-specific. The enhancements or changes may be provided by the owner, obtained from the original supplier on a per diem basis, or obtained from another firm that has experience with the application and the computer.

The owner may integrate his own system. This, however, requires a staff of engineers and programmers. Also required is a facility to stage the system for integration, software development and testing. The owner also must establish standards for documentation and their development. Future support can be expensive or impossible without proper documentation.

PROCUREMENT DIFFERENCES

Data Processing

Data processing systems should be procured based on available software, unless all software is to be developed in-house (a costly proposition). Software generally is not portable from computer to computer, except within one manufacturer's line of equipment specifically designed to be compatible. The different computers may have the same language available (e.g., COBOL). However, programs are written to work with specific utility programs, which are specific to a computer. They are standard routines used to assist in the operation of the computer. This includes conversion, sorting, printout and VDT display routines.

Procuring a data processing system should therefore follow these steps:

1. Perform a requirements analysis.
2. Evaluate available software.
3. Select the software that best meets the needs.
4. Evaluate the computer(s) that will accommodate the software.
5. Select the computer that meets the needs.
6. Program the applications for which standard software is not available on the selected computer.

Process Control

Process control systems should be procured based on a system design and specifications. Each PC application is unique. Selecting a vendor before the specific needs and objectives have been identified and documented may result in unacceptable compromises. The approach used to select software in data processing applications must be used to select the complete system in process control applications:

1. Establish needs and objectives.
2. Prepare preliminary design.
3. Evaluate available systems/vendors.
4. Prepare final design, maximizing the capability of selected equipment/system(s).

All procurements involve compromises. The buyer must be aware of the compromises he makes when he procures a software package or a control system. The detailed requirements definition is the base line for this determination.

SUMMARY

This chapter compared electronic data processing and process control. They are not identical. The following points summarize the information presented in this chapter:

1. The computer equipment may be similar, but serve different functions. The I/O subsystems and the operator interface differ greatly. Relevant experience is required to configure safe, reliable and appropriate systems.
2. The software systems differ greatly. Application requirements require different features and capabilities.
3. The programmer must have experience both with the application and the software used. This means that different people are involved.
4. The method of system selection is similar but not identical. One should know in greatest practical detail what is needed.
5. The vendors are characteristically different. Their products and support functions are different.
6. The technical background and experience of the people in system definition, design and implementation are quite different. Three areas of expertise are involved: application, software technology and scientific principles.

REFERENCES

1. Cerullo, M. J. "Developing Sophisticated Computer Applications," *J. Systems Management* (January 1980).
2. Feldman, J. A. "Programming Languages," *Scientific Am.* (1980).

BIBLIOGRAPHY

1. Copeland, J. R. "Process Control: It Isn't EDP!" paper presented at the Instrument Society of America's Industry Oriented Conference and Exhibit, Milwaukee, WI, October 6-9, 1975.
2. Edwards, E., and F. P. Lees, Eds. *The Human Operator In Process Control* (London, England: Halsted Press, 1974).
3. Knudsen, D. I., and R. C. Manross. "Computer Control of Wastewater Treatment Plants—It Takes Knowledge and Commitment," paper presented

at the Water Pollution Control Federation 51st Annual Conference, Anaheim, CA, October 4, 1978.
4. Livingston, W. L. "Hold the Control Loop Designs: Do the Systems Engineering First," *Control Eng.* 79 (January 1981).
5. Sheridan, T. B. "Computer Control and Human Alienation," *Technol. Rev.* 61 (October 1980).
6. Manning, A. W., and D. E. Dobs. "Handbook for Automation of Activated Sludge Treatment Plants," U.S. Environmental Protection Agency, Report No. EPA-600/8-80-028 Cincinnati, OH (1980).
7. Williams, T. J., and E. J. Kompass, Eds. *Man-Machine Interfaces for Industrial Control* (Barington, IL: Control Engineering, 1980).
8. Carroll, R. E. "Guidelines for the Design of Man-Machine Interfaces for Process Control," Laboratory for Applied Industrial Control, Purdue University, West Lafayette, IN (1978).
9. Kanter, J. *Management Guide to Computer System Selection and Use* (Englewood Cliffs, NJ: Prentice-Hall, Inc., 1970).
10. Kepner, C. H., and B. B. Tregoe. *The Rational Manager—A Systematic Approach to Problem Solving and Decision Making* (New York: McGraw-Hill Book Co., 1965).
11. Wooldridge, S. *Software Selections* (Auerback Publishers, 1973).
12. Kanter, J. *Management Oriented Management Information Systems* (Englewood Cliffs, NJ: Prentice-Hall, Inc., 1976).

CHAPTER 3

CONTROL SYSTEM HARDWARE

Paul V. Manns, PE

INTRODUCTION

The selection of hardware for a successful computer control system requires a thorough knowledge of the process, instrumentation, computer equipment, telemetry, panels and site planning. This chapter provides an overview of system design considerations and available capabilities and explains many of the pitfalls common to these systems.

This chapter covers basic computer hardware. It covers other system components such as computer peripheral, process input/output (I/O), instrumentation and panels. There are several system configurations available and these are related to preexisting control systems, the exposure to control failure and relative control benefits.

Computer

A process control computer executes the instructions that determine how a process is monitored and controlled. The instructions are usually found on disk files. They operate on dynamic data originating from field instrumentation and the process operators. The computer scales instrument data, compares it with tables, generates displays for the operators, stores information in files for reports and determines which control actions are required from mathematical control programs. Operator requests for displays and reports are honored by transferring data stored on the disk to the printer or operator console.

Computer Peripheral

The peripheral equipment is attached to the computer and provides operator control, storage of information and printed results of computer processing. Examples of peripheral equipment include the operator's console, mass storage disks, magnetic tape units, diskettes and printers. Smaller computer systems, especially microcomputer-based systems, integrate all but the printing functions into a package that resembles a terminal more than a traditional computer.

Process Interface

The process interface provides the sensory capability for the process computer. A specialized I/O channel, or remote multiplexer, samples all the instrumentation and transfers this information to the computer memory. Commands are transferred from the memory to control interface units, which are designed to change valves, motors, speed controls or other equipment that physically controls a process.

Telemetry Equipment

Telemetry equipment provides a means to observe and control conditions at a location remote from the computer. Typical applications provide central monitoring for electrical, gas, water and waste water collection networks. Other applications include pipeline control and pollution monitoring.

Remote process information is converted to an electrical signal, which is transmitted as a tone or set of pulses over a telephone line, wire carrier current or radio to a central monitoring point. Most applications provide monitoring only, but remote control is common for pipeline and electrical distribution networks.

Panels and Terminations

Control panels permit an operator to concurrently observe interrelated process conditions and make appropriate changes. In a noncomputerized control system there is typically a centralized panel with automatic analog and sequence controllers. All plant operations are controlled at this point. A computerized plant is more likely to have panels distributed throughout and have less sophisticated control on each. These panels are provided as a backup to the computer.

Field Instrumentation and Controllers

The field instrumentation measures the parameters and status of the process. Typical process measurements include temperature, pressure, level, flow, density and position. Each type of process will require specialty instruments to measure critical parameters such as moisture and thickness for paper, and suspended solids and dissolved oxygen for waste water. pH, color and elemental composition analyzers are used in other industries.

Status switches are used to indicate alarm conditions, mechanical limits or equipment running status. The switches may be actuated mechanically or may be part of an electrical relay control device.

Control devices that physically regulate the process include valves, pumps, blower conveyors and mechanical actuators. They are activated by electrical signals, which, in turn, control electrical, hydraulic and pneumatic actuators.

PROCESS CONTROL COMPUTERS

General

The process control (PC) computer is surprisingly small when compared with the other process control elements. In a large system it is typically mounted in a 19-inch equipment rack and rarely exceeds 18 inches in height. The computer in a small system may be a printed circuit card or even one silicon chip buried in a rack of process input/output cards.

The computer has three distinct sections that work together to execute the control system software. These are the control logic, memory and I/O control (Figure 1) and are interconnected with a data and control pathway, or bus.

The control logic controls the sequence of instructions executed by the computer. It also interprets these instructions and performs the arithmetic, comparison and I/O initiation functions required to execute a control program, respond to a human operator or to transfer a graphic display from mass storage to the operator display.

44   PROCESS CONTROL COMPUTER SYSTEMS

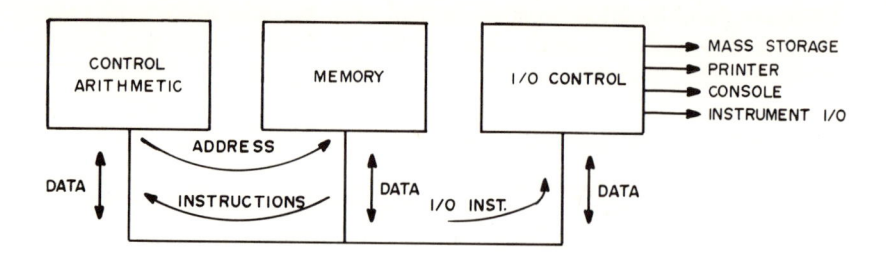

Figure 1. Basic computer structure.

The memory provides a medium to store instruction and serves as a medium to stage information as it is transferred among the mass storage media, operator display and field equipment.

I/O control executes the transfer of information between the computer memory and the mass storage operator display and field equipment under direction of the control section of the computer.

Computer capability is measured in terms of its instruction execution rates, memory capacity and input/output capacity. These three characteristics form the basis that separates the simple single-loop microcomputer, PID (proportional, integral and derivative) controller or programmable sequence controller from a computer that controls an entire industrial complex.

Modern process control systems have evolved into a system of connected computers in which each remote computer handles a portion of the total process. The central computer solicits data from and passes optimization information to the remote computers. This organization is found not only in large petrochemical complexes but even within some industrial robots.

Microcomputers

The microcomputer normally is thought of as a "computer on a chip." However, there are variations. A simple microcomputer system found in a calculator, digital PID controller or low-cost digital telemetry equipment may have the controller, memory and I/O on one chip. Larger systems have a single control and arithmetic chip, but the I/O and memory are added on as required. The computer seldom occupies more than one or two PC cards. In fact, it may be difficult to determine whether the peripheral controllers and process I/O are integrated into the same package. Microcomputer-based control systems seldom control more than 40 loops and are used in small systems or as part of a larger distributed control system.

Minicomputers

The mainstay of the digital process control industry has been the minicomputer. It is packaged as a complete unit and typically has 10 times the program execution power of the microcomputer. The computer is typically mounted in an equipment rack with the peripheral control, process I/O, disk and communications equipment. The system can handle several thousand points with the aid of remote multiplexers or remote microcomputer process controllers.

The minicomputer-based system can support several color operator consoles, high-speed printers and magnetic tape while performing its primary process control functions.

Large Computers

Large computers always have been sold as a complete data processing package by the large computer companies. Process control systems houses have found the large systems inconvenient and uneconomical to incorporate into a control system.

Special-Purpose Computers

The microcomputer has spawned a great number of products because of its low-cost versatility. Two of these units are of special interest to control system users.

Programmable Logic Controller

The programmable logic controller (PLC) is especially designed to replace panels designed with relays. The control language supplied with these units emulates ladder diagrams and can be understood and programmed by a good electrician. Some PLCs also have analog control capability and increasingly approach the versatility of a microcomputer-based process controller.

Remote Telemetry Unit

The microcomputer-based remote telemetry unit (RTU) not only has the power to transmit data and accept commands, it also can replace the control functions normally designed with relays in a pump station panel.

COMPUTER PERIPHERAL

Operator Consoles

Operator consoles can range from a simple keyboard-printer terminal to multiple color graphic video display terminal (VDT) displays with light pen, menu and control selection. The color graphic VDT display has become common in systems with as few as 200 points. Color provides the capability to present a very detailed display with the most important information highlighted for instant recognition. Graphics permit the process information to be presented with comparative bar charts or as a labeled schematic representation of the process. Black and white alphanumeric displays are found on very small systems or as backup operator access to remote units of a distributed system.

Keyboards provide operator access to the system and can have several components. The standard typewriter format keyboard permits the operator to select items on the display by name. Some systems do not require this keyboard because they employ menu-type selections requiring no alphabetic entries. A function keyboard provides functions similar to those found on a conventional panel and permits display control, valve and motor control, setpoint control and other functions that may be unique to the system. Many systems provide a numeric keypad with a calculator number arrangement to permit rapid numeric value entries.

Other human factors designed into a quality console include tinted displays to enhance display contrast and a tilted back-frosted glass display to eliminate glare and reflections. A plantwide voice communications system permits the operator to coordinate with instrument maintenance people for calibration and instrument availability information. An audible alarm annunciator is often provided that can be acknowledged from the function keyboard.

Printers

Printers provide hardcopy records of alarm events, process summary reports and, in some cases, a graphic display. The most important consideration for a printer in this application is reliability.

CONTROL SYSTEM HARDWARE   47

The printers are typically 100 to 150 characters per second (cps) wire matrix serial input units that print 132 characters per line on 17.5-inch-wide fanfold paper. High-speed line printers are rarely required on a process control system, but the inexpensive 10–30 cps printers are appropriate only for programmers terminals or for extremely small data logging systems.

Disk Systems

Disk units supply both permanent and supplemental data and software storage for a computer control system. Three classes of disk storage are available today, including the diskette (or floppy disk), the removable media disk and the fixed media disk.

The diskette drive unit is commonly used as the only data storage media in hobby and small business systems. Diskette drives are not designed to provide the continuous online service required in a control system so are used only to load software into permanent storage or for providing long-term historical storage on diskette media. The 8-inch units provide 200,000–1.0 million bytes or characters of storage per diskette.

The removable media disk drive units can be found on some of the very large (more than 5000 points) control systems. These units can store 30–500 million bytes of information per drive. Several disk packs with control system software can be stored to back up the online pack. Removable media disks are slow compared to fixed media disks so that the large control system usually also has a fixed media or fixed head disk to store data requiring frequent access.

Fixed media disks include both fixed and movable-head disks. Fixed head disk units provide fast data access but are usually limited to 20 million bytes capacity and cost more per byte storage than other disks. The older large dual central computer type systems used fixed head disks to help increase the scan rate, but newer distributed systems seldom require fixed head storage.

Magnetic Tape Drives

Magnetic tape drives are used to transfer software and to archive data. Magnetic tape can store more information per cubic foot of storage space and do it more inexpensively than any other media.

If archive data is to be processed on a data processing system, the tape recording format and density must match on both computer systems.

Most tape transports provide switch selection between two or more formats. High-performance tape systems are not necessary on control systems, and the tape cartridge (similar to a Phillips cassette) may satisfy program loading and data storage requirements for small (less than 200 points) systems.

PROCESS INTERFACE I/O

There are two types of process information available to a control system: digital information and analog information.

Digital information can be visualized as an event that is either occurring or not occurring, such as:

1. The hydraulic pump is running (or off).
2. The pressure is too high (or ok).
3. The valve is open (or not open).

This type of event closes (or opens) the contacts of a mechanical or electronic switch and the computer senses the position of the switch.

Digital control or output from a control system closes a relay contact, which, in turn, may start a motor or actuate a pneumatic valve.

Analog information includes value data such as:

1. How fast is the pump running?
2. What is the pressure?
3. How far is the valve open?

Analog information is transmitted from the sensing instrument as a voltage or current that ranges between two limits such as the standard 4-20 mamps or 1-5 V. The I/O equipment converts the analog signal to a digital word (8-16 bits) using an analog to digital (A/D) converter. This format then permits the computer to process the analog information as a convenient digital number.

Analog control from a computer can be incremental or positional. Incremental control is typical where a reversible motor-driven actuator moves a control device. The computer closes the control contact for the appropriate movement until the process correction is complete. A digital to analog conversion is required for positional control, and the position signal is normally presented to the actuator as a standard current or voltage. This type of interface is typical of speed controllers or servo control.

Multiplexing and Remote Processing

Process instrumentation is located at the process sites and thus poses an information transfer problem in a spread-out system. It would be possible to wire all instrumentation to the central computer but this can represent a great investment in conduit, wire and labor. Multiplexing permits the conversion of all information to digital form at remote sites throughout the plant, and transmission of all information to the central computer on a serial data highway. Typically, this data highway is a single pair of wires that requires conduit only for mechanical protection.

Multiplexing can be accomplished with a hardwired multiplexer or with a microcomputer that not only will multiplex data but will perform scaling and basic control functions that continue even with the loss of the central computer.

COMPUTER CONTROL SYSTEMS

Centralized Systems with Integral I/O

The computer control system with self-contained I/O (Figure 2) is cost-effective in situations in which all instrumentation is available at a single location without extensive field wiring. These include small systems with less than 100 points and systems adapted to an existing centralized control panel.

The small system often lacks such amenities as process graphic displays, a report printer or dual backup capability found in the more elaborate systems. The larger systems of this type are not often found today but were common during the period when large centralized panel systems were being converted to computer control.

Centralized Computer with Distributed I/O

Computer systems installed in plants with no centralized panel avoid the problem of expensive wiring costs by installing remote I/O units throughout the plant (Figure 3). The cable from the I/O multiplexer unit to the central computer requires less-expensive conduit and a single wire pull. These systems have no central backup panel so reliability becomes extremely important. It is common to configure redundancy wherever

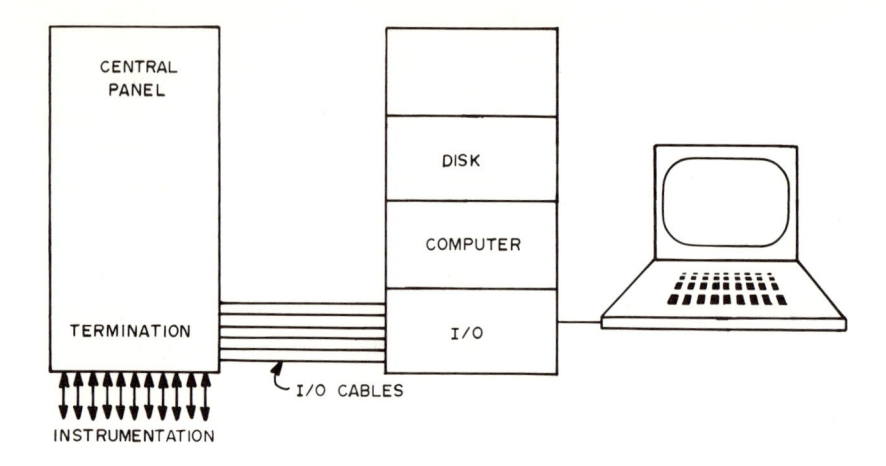

Figure 2. Centralized system with integral I/O.

possible, including two computers, two sets of multiplexer cables and uninterruptible power. The area panel computer interface is designed so that all computer control is accomplished with momentary contacts to reduce the risk of a computer failure, which would cause a serious process upset.

Distributed Intelligence Systems

The advent of low-cost computer power in the form of the microcomputer has led to the development of hierarchical systems in which microcomputer systems placed close to the process provide basic regulatory control (Figure 4), while the central computer assumes a supervisory role in the system. The major advantage of this approach is that control failures are localized and a complete system failure is highly unlikely.

The central computer system is less likely to require an uninterruptible power system (UPS), fast fixed-head disks or a dual configuration because essential control does not depend on the central computer. The central system typically provides color graphic process displays, the ability to modify control executed in the remote process computers, report alarms and print operating reports. Often the central system will coordinate power demand and offer plantwide control optimization algorithms, which are important but not critical.

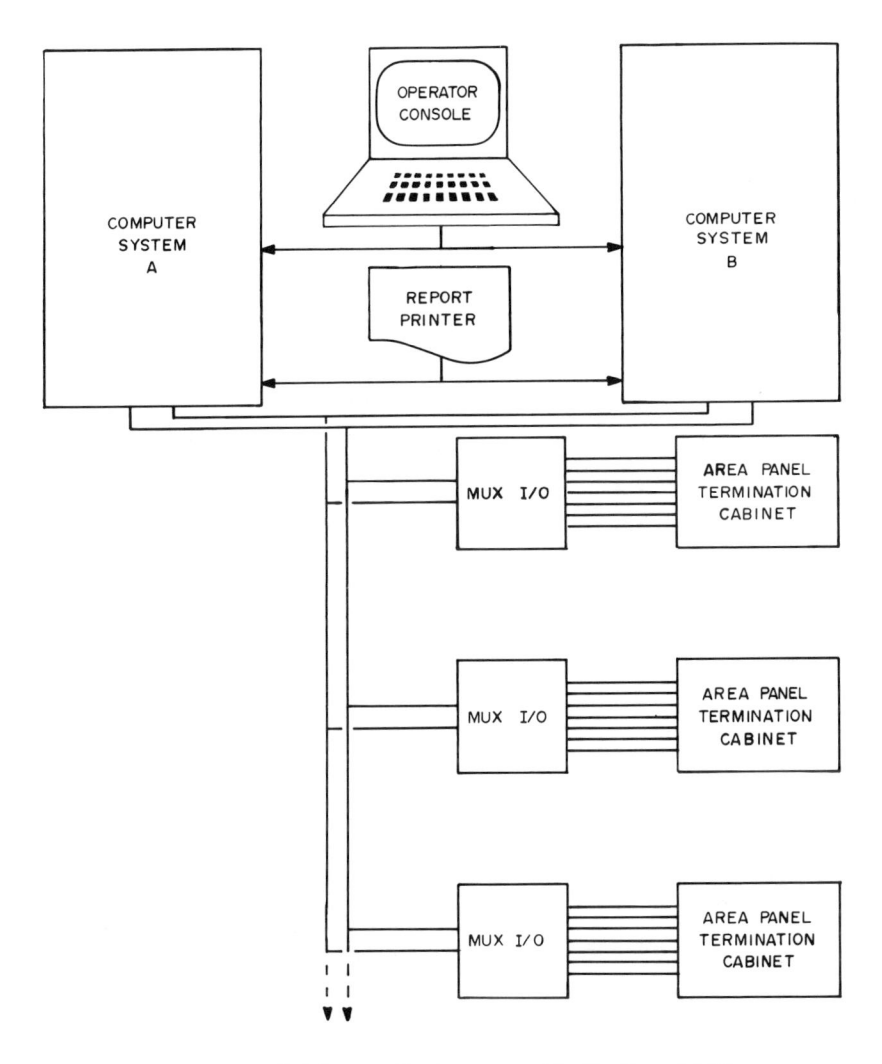

Figure 3. Dual centralized system with distributed I/O.

The process computers in Figure 4 are located near the process being controlled and are often microcomputer based. Control algorithms can be programs stored in read only memory (ROM), but the control configuration or "softwiring" and tuning can be changed either from the central computer or from a locally connected terminal.

Many remote process computers originally were developed as very small, stand-alone control systems and have the provision for operator

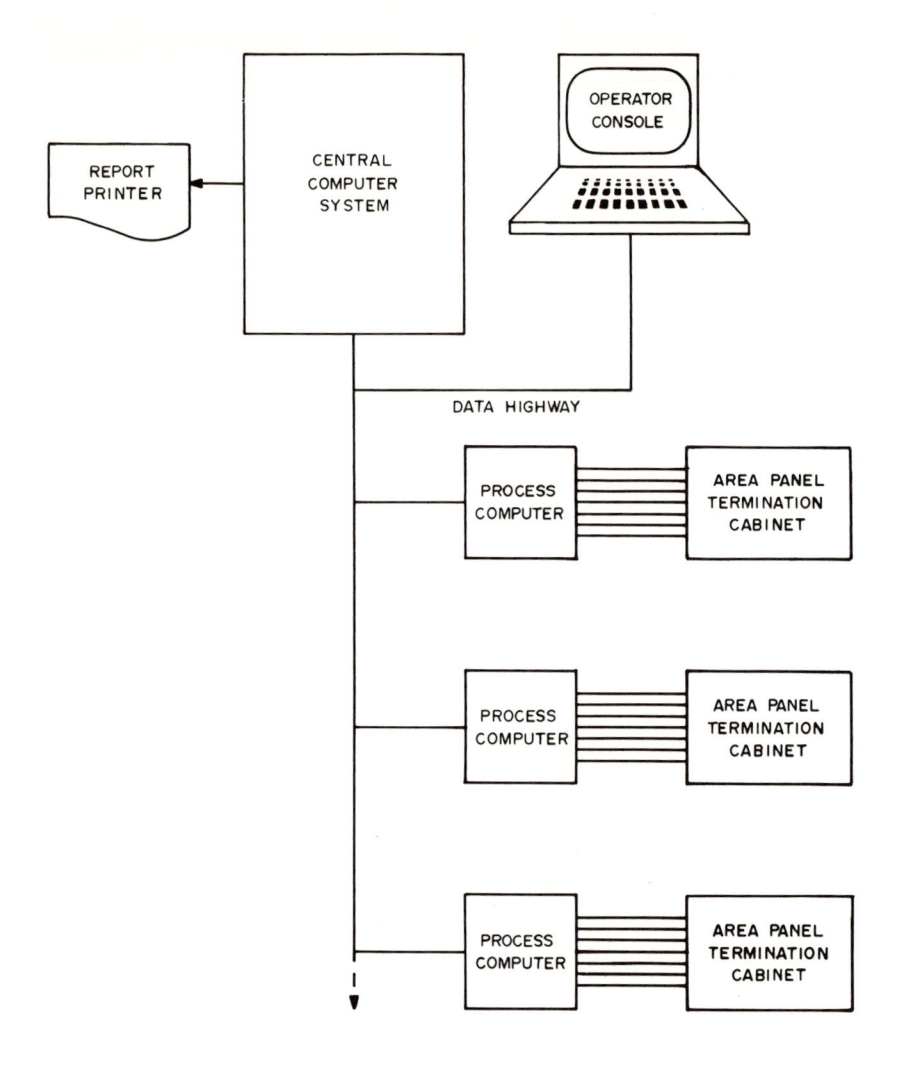

Figure 4. Distributed process control system.

control. The operator interface is typically a VDT display, with bar graphs showing process and controller status. Overview and detailed loop displays also are available. The keyboards are a waterproof membrane design with numeric, menu selection and control functions.

The link that ties the computer system together is referred to as a data highway or, more generally, as a local area network (LAN). It is not practical to mix different manufacturers in these systems because each has its unique electrical, protocol and data structure interface.

Data Logging Systems

Small processes with existing analog panel control may only require an operating log with trending and alarm recording. Equipment can vary from a simple strip printer recording analog meter information to a distributed I/O system with process graphics. These systems, while difficult to justify because there are no manpower or process cost savings attributable to them, do provide centralized process information. They may or may not utilize remote multiplexing, usually depending on the location of the existing panels.

Hybrid Systems, Interlocks and Packages

The hybrid system is an approach to control system reliability that uses the computer system to regulate the setpoint of analog controllers. The system is conceptually similar to the distributed intelligence systems, in which the analog controllers are used instead of the remote process computers. The concept was popular before 1977, but the lower cost and added flexibility of the remote digital computer has relegated this form of control redundancy to unstable processes, which must have some form of regulation at all times.

A process that could become dangerous or damage equipment should have hardwire backup interlocks to prevent the hazard should the computer fail. Package process systems usually include controls designed expressly for the package, and the computer interface should only be used to start, stop and monitor the package equipment.

PANELS AND TERMINATIONS

Control panels with their collection of switches, indicator lights, analog indicators and controllers and trend charts have been the traditional process control stations for more than half a century. Production panels are normally supplied with packaged process equipment, but larger areawide and plantwide panels are individually designed and built to specification. Simplified panels often are included with a computer control system for manual backup control. Additionally, panels often serve as a location for terminating and concentrating signal wiring from various locations.

54 PROCESS CONTROL COMPUTER SYSTEMS

Panel Systems

Local control stations are small, predesigned units supplied with valves, motors, conveyors and other process control equipment. They usually control and maintain the equipment within visual contact of the operator. These control stations may be as simple as a start/stop switch or may be an organized panel of several switches and controllers for a small package process unit. In many installations these are intended for maintenance only because control for an entire process is concentrated on another panel or in a computer.

Area panels, sometimes known as process panels, are custom designed to operate an entire process or area of the plant. A single operator can view the operation of the system and make the necessary adjustments. Process systems with central panels or computers may not have these panels in all process areas if backup control is not critical and can be handled from the local control stations.

Centralized panels are designed to cover an entire plant or the telemetry system, or both. Control systems employing computers seldom require the centralized panel.

Analog Control

Analog control equipment includes all noncomputer equipment designed to regulate a process in a continuous mode. The basis of analog feedback control is a device that compares the desired operating conditions (setpoint) with the actual operating conditions (process variable) and sends a new control device position to correct the process to the setpoint. Rate or proportional controllers are feedforward systems that adjust process equipment based on a mathematical computation rather than on empirical feedback.

A description of all available control devices of this nature would fill a book several times this size; however, they all fall into three broad classifications.

1. **Mechanical regulators** include fluid-filled temperature-regulating devices, pressure-controlled dampers, pressure-regulating valves and mechanical-speed governors. These controls are designed as part of the equipment supplied and are difficult to adapt to remote adjustment.

2. **Pneumatic control** often is found in pneumatically driven process equipment. Pneumatic functions operate from a supply of air. Air pres-

CONTROL SYSTEM HARDWARE

sure transmits all control signals standardized within the range of 5 to 15 psi. Pneumatic control devices include addition, multiplication, root extraction and setpoint PID control. These devices can interface directly to pneumatically driven servomechanisms without an electrical to pneumatic signal converter.

Pneumatic control has been utilized in large systems but is somewhat slow, difficult to maintain, subject to moisture and does not have the variety of instrumentation available as do electrical control systems. Electrical to pneumatic converters are available to interface standard electrical analog signals to standard pneumatic control pressures.

3. Electrical analog control devices include signal conditioners, setpoint controllers, manual loading stations, indicators and strip chart recorders. Most analog equipment has been standardized to interface to a 4- to 20-mA or a 1- to 5-volt analog signal. These devices have many suppliers, a rich array of compatible instrumentation and control devices, and standardized computer interface equipment. Most control panels have electronic setpoint regulators, instrument display equipment and strip chart recorders mounted for operator control and display. Signal conditioners such as square root extractors, adders, etc., are mounted behind the panel or near the instrument.

Switches and Indicators

Two general types of motor control switches are in general use: (1) the rotary on/off or hand/off/automatic-type switch and the start/stop momentary contact-type control.

The rotary switches provide the lowest cost control switch and are self-indicating. If no computer control of the device is required and its running status is either obvious or presents no safety hazard, then the rotary on/off switch is an appropriate choice.

The start/stop momentary contact-type control is more expensive because it requires lockup relays to keep the device running and running status indicators if running is not visually obvious from the control switch; however, this arrangement has several advantages, especially when used with a computer control system:

1. It has superior operator control.
2. The device will not restart after power failure recovery.
3. Computer/manual transfers are "bumpless."
4. The computer is not required to maintain the running relay in computer mode.

Sequence Control

Sequence controllers are used to start equipment when a definite time and event sequence is required. Traditionally, this has been implemented with cam and switch arrangements or with timers and relays. Microcomputer-driven PLCs are now available that provide all timing and relay functions in software. Either type of sequence control can be efficiently interfaced to a computer with the equivalent of startup and shutdown controls from the computer and with running and alarm statuses returned to the computer.

Panel Layout

The layout of a control panel is very important from the human factor standpoint. The panel controls should be placed so they can be associated with the geographic location of the controlled devices or associated with the process flow. All controls and indicators associated with each other should be grouped together. Computer/manual switches should be placed adjacent to the associated manual controls. If a single computer manual switch is associated with a group of manual controls, this association also should be clear. Careful panel design will minimize operator error by clearly showing what is controlled, the effect of that control and what controls and indications are associated with each other.

TELEMETRY AND COMMUNICATIONS

Often the process control requirement does not stay within the confines of the plant, especially in the case of the utility companies. The distribution network may require more monitoring and control than the physical process plant itself. For metering and failures recovery, it may be necessary to coordinate multiple sources contributing to the system. Three principal methods of communication are telephone lines, radio and carrier current.

Telephone Systems

Two general types of telephone lines are available and are classified as leased lines and switched lines. The leased lines of interest are the 1000-type and 3000-type lines. Switched lines use the normal dialup network.

Leased Lines

Much of the older telemetry equipment required a dc control signal to operate a relay at the remote site. The telephone company provided a metallic pair of wires between both sites so direct dc control could be accomplished. The telephone company used the same type of low-speed line for teletype signals and classified this grade of line as a 1000-type line. Many systems of this type are still in use today, but the telephone company has found much more efficient means to transmit information and no longer guarantees the availability of dc-type 1000 lines. The modern 3000-type line is similar to a voice line but has a specification more suited to tone and digital transmission equipment.

Most leased lines incur a monthly charge and include a flat monthly fee for a line to the telephone exchange plus a mileage charge between exchanges. Line charges can be reduced significantly by specifying multidrop configurations, which link up to 20 stations to a single telephone line.

Both duplex (simultaneously both ways), or four-wire, and half duplex (alternately both ways), or two-wire, lines are available, depending on the requirements of the telemetry terminal equipment. Duplex lines usually cost twice the half duplex rate and should be avoided except where the required amount of information cannot be handled within time constraints imposed by the half duplex lines.

Switched Lines

Switched lines can be used to advantage when remote information or control is only required on an exception basis, such as a remote alarm, a daily meter reading or a twice daily mode shift. Switched lines are part of the everyday dialed voice system and usually cost less than leased lines when an entire metropolitan area must be covered or where voice telephones are installed for other purposes.

Radio Systems

Two general types of radio equipment are commonly found in telemetry systems. These are UHF radio and microwave radio.

UHF Radio

UHF radio is an economically viable alternative to telephone communications where five or more sites are located remote from existing telephone lines. Installation of all radio equipment requires a Federal Communications Commission (FCC) license, antenna equipment, special radio control equipment, a power source capable of running the radio equipment and a base station capable of reaching all remote units. In most localities, the telephone company will assess an installation charge exceeding the cost of a UHF radio station to extend a line more than 2000 feet.

Remote UHF sites are only used with digital telemetry equipment because it must be programmed to respond to polling in the receive mode, activate the transmitter, quickly transmit its digitally encoded message and revert to receive mode.

Microwave Radio

Microwave radio equipment can transmit 20–100 times more information than telephone or UHF systems, but signal transmission is highly directional. The primary application of microwave equipment includes pipeline control, coordination of multiple computer systems or use for data trunks, which combine information from several telephone trunks.

Tone Equipment

The standard method for transmitting analog or digital information over telephone lines has been the utilization of separate tones for each data channel. This is known as frequency division multiplexing. Most of the existing telemetry equipment today is this type.

A single-tone channel can handle up to 32 digital points with the use of a scanner, which sequentially transmits the status of all points. The receive scanner must be synchronized to the transmitter to properly sort each point for display on panel status lights. Analog information is transmitted using time duration methods. The tone is keyed by an electrical contact, which is physically moved by a mechanical cam connected to the process variable. Received information typically is displayed on chart recorders.

Tone equipment is most suited to applications in which fewer than 10 signals must be transmitted to a panel display. It becomes more susceptible to interference as more channels are added. There is no error correction capability, so remote control is a tricky proposition at best.

Status information can be connected directly to a computer because a status signal at the receiver is typically a relay contact. Some computer systems provide a time duration analog interface option, but most systems require a converter to change time duration inputs into 4–20 mA signals.

Digital Equipment

The marriage of digital data communications techniques, with its sophisticated protocols and error correction techniques, to telemetry technology has produced what is now called digital telemetry. The remote equipment has evolved from a simple scanner to a microprocessor unit capable of analog to digital conversion, polled protocols and some local automatic control functions.

Typically, a remote digital station will accept a limited combination of analog and digital instrumentation and control. An integral modem converts the digital information to a frequency-modulated tone suitable for transmitting information at 10–120 cps. Of these units, 10–20 can be placed on a single telephone line but must be coordinated by a master unit (which may be the process control computer), which sequentially requests information from each of the units. There are digital units designed to dial and answer switched telephone lines or to operate UHF radio stations.

The versatility of microprocessor-based equipment is both an advantage and a disadvantage because of the almost unlimited combinations available. The protocol that defines the data format and error recovery is not standardized. Therefore, equipment from different manufacturers is not interchangeable. Even single suppliers may have different types of equipment that are not compatible. One should not mix suppliers of digital equipment without the counsel of an expert in this area.

The computer hardware interface to digital telemetry equipment is simple compared to the tone telemetry interface. However, the protocol and software handlers required usually preclude mixing suppliers of process control computers and telemetry equipment without expert attention to the software interface requirements.

FIELD INSTRUMENTATION

Without measurement or event detection there can be no control. There must be a means to determine whether the process is within tolerance and, if not, what correction is required.

The instrumentation measuring basic properties of pressure, level flow

60 PROCESS CONTROL COMPUTER SYSTEMS

and temperature are well developed and reliable when properly applied. Other properties such as density, turbidity, viscosity, etc., are measured with instrumentation, which is very application dependent and specialized. Analytical instruments are used to measure physical or chemical properties of fluids or materials and typically employ balanced chemical cells, gas chromatography, spectroscopy or reaction displacement devices.

The field of instrumentation covers a broad spectrum in the literature, so this section will survey only the instrumentation most commonly applied to process control.

Pressure Measurements

The most common pressure measuring instruments applied to process control include the diaphragm gauge, the bellows and the strain gauge. Pressure instrumentation is selected on the basis of measurement range, price versus accuracy, and sensitivity to damage from pressure surges, clogging or corrosion.

Diaphragm pressure gauges utilize the deflection of a sealed metal diaphragm. The diaphragm forms one plate of a capacitor and the electrical pressure signal is derived from a capacitance bridge. Special versions are available with neoprene seals to measure the pressure of corrosive liquids. Diaphragm gauges are commonly used in mid-range pressures from 1 to 5000 psi.

The bellows pressure gauge resembles a brass accordion internally and is applied where pressures do not exceed 100 psi. The bellows travels 5–10% of its length over its pressure range and provides mechanical movement for indicating transmitters. Electrical signals are derived from reluctance or Hall effect displacement transducers.

The strain gauge utilizes the principle of electrical conductance variations as a result of mechanical strain and takes advantage of metal deformation, piezoelectric effects, ceramic transducers or semiconductor transducers. Strain gauges are well suited for operation in the high pressure ranges.

Level Measurements

Level measurements are commonly required for pump and valve control from or to an open tank or container. Most level instruments are suited to liquid level, although solids level can be measured if the surface is reasonably flat or if weight and level correspond.

The bubbler is a common level measurement device. It includes an air compressor, a tube that reaches to the bottom of the liquid and an air pressure gauge. A constant flow of air is supplied to the tube so that a small amount of air is "bubbled" from the bottom of the tube. The resulting back pressure is directly proportional to the liquid level. The pressure signal is scaled in terms of liquid level. The bubbler is accurate, relatively free of fouling and has found wide use in a variety of applications. The bubbler is relatively expensive to install and operate.

Another approach is the pressure gauge. It can be installed at the bottom of a tank and the signal scaled in terms of level. Access is often impossible, and plugging from sedimentation can be a problem unless the gauge inlet is designed properly. The level in a solids bin is often determined by weighing the entire bin with a strain gauge. The signal is scaled for level and the empty bin weight is used as a zero offset.

A float installed in a stilling well is mechanically cabled to an indicating drum and a counterweight. The drum rotates as the level changes and provides a visual level indication. The electrical signal is derived from a potentiometer or encoder mechanically coupled to the drum. The capacitance probe detects the change in dielectric constant along a probe as a liquid or solid moves along the probe.

The ultrasonic level detectors operate on the sonar principle and can give erratic signals unless properly installed and adjusted to reject reflection from the side of the container or tank.

Flow Measurement

Fluid flow measurement instruments become part of the conduit and thus have many variations, depending on the fluid, required accuracy, open channel or closed pipe, solids or gases suspended in the fluid, turbulence and temperature of the fluid. Flowmeters lose accuracy at the low end of their range and some will not operate below a minimum threshold. All factors must be considered for proper flow instrument selection.

Differential pressure flowmeters depend on the deliberate introduction of a controlled pressure drop in the pipe using an orifice plate, a venturi or an annubar. The pressure drop is proportional to the square of the flow. This class of flowmeter works well in clean fluids but is susceptible to plugging, and the low flow inaccuracy is degraded by the square law response.

Related to the differential pressure flowmeters are the open channel, weir and parshall flume flow measurements. The weir is equivalent to an orifice plate and the flume is equivalent to the venturi. Flow is translated to level in the weir or flume and is scaled in terms of flow.

Mechanical flow measurements include the turbine flowmeter, the positive displacement pump and piston-type flowmeters. Turbine flowmeters rotate in proportion to flow, and the flow signal is a variable-frequency pulse rate derived from a magnetic pickup. A positive displacement pump provides flow information by virtue of the known flow from each stroke. The rotation piston or disk is commonly used to meter household water use and employs a piston that "wobbles" at a rate determined by the water flow. Emulative flow is accumulated on a local or remote pulse counter.

The electromagnetic flowmeter or magmeter generates a magnetic field through the measured liquid. As the liquid moves or flows through the field a voltage is generated. This is sensed by electrodes placed perpendicular to the magnetic field. To function properly, the liquid must conduct electricity and the pipe must be full at all times. Magmeters can be supplied with ultrasonic electrode cleaners to assure conductive contact with the fluid. Magmeters are quite expensive but provide high accuracy and reliability.

Sonic doppler flowmeters operate on the principle that a moving liquid will change the travel time of sound through the liquid. Transducers can be applied external to the pipe. Accuracy approaches magmeter standards but transducers are sensitive to turbulence and bubbles in the liquid.

Temperature Measurement

The thermocouple operates on the principle that a voltage generated across the junction of two dissimilar metals is related to the temperature of the junction. Resistance thermometer devices (RTD) operate on the principle that metals change resistance with temperature. A bridge circuit is used to detect the resistance change. Thermistors are similar to RTDs but are composed of semiconductor material, which is much more temperature sensitive (and less accurate) than metal RTDs.

The radiation pyrometer detects the temperature of a radiating body by focusing radiant energy on a thermocouple or semiconductor temperature sensor. High-temperature optical units compare the radiating body color with a filament of known temperature.

Density and Suspended Solids

Density can be measured by weighing a known volume or by detecting nuclear radiation absorption. Suspended solids are measured using optical absorption or scattering methods. Suspended solids instrumentation is subject to fouling in many applications, and careful research into the suitability of any instrument in this area is required.

Analytical Instrumentation

Analytical instrumentation measures physical or chemical characteristics of a material or compound. These instruments are very specialized and each industry has unique requirements. Some general principles and the more common applications are discussed here.

pH measurement is accomplished using a permeable electrochemical cell balanced with a reference cell. The interface between the measured liquid and the measuring cell is referred to as an electrode and is composed of porous glass or antimony. The type of cell must be matched to condition in the measured liquid. Glass is more fragile and antimony requires more maintenance.

Dissolved oxygen can be measured in a liquid cell and is widely used to control oxygenation processes. Similar cells are available to measure other dissolved gases such as chlorine. Solid cells are available to measure flue gas oxygen to control flame efficiency.

Chromatography is used to detect the amount of a selected compound in a mixture. The components are converted to a gaseous phase if required and separated by a diffusion or adsorption media. The relative amount of a compound is detected from a physical characteristic such as thermal conductivity, specific heat, flame ionization or mass. Successful application of this equipment depends on the ability to maintain the instrument in the pressure of contamination.

Spectrum analyzers detect the presence of chemical elements in gaseous form by measuring light wavelength absorption or flame color emission. Color separation and detection is done with prisms and photodetectors.

Field Status Input

The discrete input signals to the computer notify the operator when limits have been met or exceeded, whether equipment is operating or whether a manual switch has transferred control from automatic to manual mode. Valve and gate travel limits are used to determine process routing or abnormal modulating valve operation. Virtually all motor controllers have an auxiliary running contact to control a running light and/or a computer input. All computer/manual switches should have a contact for the computer to indicate its position. Many of the instruments described previously have a version that simply activates a switch if the process measurement is above or below a preset value.

Control Devices

Instrumentation is the sensory portion of a control system. The control devices provide a means of corrective or initiative control. Typical control devices include valves, gates, motor-driven pumps, blowers, etc., as well as servo controls.

Valves range from a simple solenoid-operated air valve to large custom-designed modulating valves. They can be classified as open/close valves or modulating valves, regardless of size. The valve actuator can be pneumatic, hydraulic or a reversible motor-driven unit with limit switches and a position transmitter. The valve is modulated with incremental motor operation in the appropriate direction. Modulated valves require a flowmeter for feedback control, preferably in front of the valve to reduce turbulence. Valve and gate control loops should be tuned to obtain desired results with a minimum of valve activity as the drive mechanisms are subject to wear.

Motor-driven process equipment can be classified as constant speed (on/off only), multiple speed or variable speed. Constant-speed motor controls simply operate a relay that operates the motor contactor in the motor control cabinet. Running contacts verify that the contactor is closed. The preferred method to interface the motor circuit from the computer is with momentary start and stop contacts. This requires that the manual motor controls operate with start and stop pushbuttons rather than a positional selector. Multiple-speed motors operate in the same manner except that an additional contact is required for each speed.

Variable-speed devices can control the motor directly with a variable frequency drive or a wound rotor drive. These devices typically require a 4–20 mA interface and require a manual loading station to convert the computer increase-decrease signals to a 4–20 mA signal. The variable-speed electric and hydraulic-speed electric and hydraulic drives also normally require a 4–20 mA control. If the computer output is incremental, then the device speed must be fed back to the computer either from a 4–20 mA tachometer or with the 4–20 mA control signal from the loading station.

Servo actuators are typically driven with a 4–20 mA positional signal. The positional feedback required is part of the servo system and may be available for positional feedback to the computer.

Analog Signal Transmission and Conversion

Standards have been established for electrical and pneumatic signal transmission. Signal converters are available for virtually any combination. Quite often it is necessary to utilize existing pneumatic valves, pressure sensors, etc., in a computer system which, of course, is electrical. The typical 5-15 psi pneumatic signal can be transformed to a 4-20 mA electrical signal, and the 5-15 psi pneumatic control can be derived from a computer-driven loading station. Electrical conversion of a 4-20 mA signal to a 1-5 V signal is accomplished with a 250-ohm resistor in series with the current loop.

Analog wiring should be shielded from electrical interference and grounded at either the signal source or the computer, but not at both places. High-current electrical power wiring should be at least 3-4 ft from analog signals where they have parallel runs.

SITE PLANNING

An important aspect of any process control system design includes providing the correct environment for the control system and its operators. Computer equipment is no longer as power hungry and sensitive to its environment as previous generations. However, site planning is still important.

Control Room

The control room must provide a suitable environment for the man-machine interface and include proper lighting, noise control and fire protection. Lighting should be designed to eliminate the reflections of fixtures on the display consoles, while still facilitating reading. The light level at the operator's console should be adjustable to strike the correct balance between background light, keyboard lighting and display contrast. The color background should be a light neutral color, such as beige or light grey for proper light diffusion and to eliminate annoying distraction or color clashes.

Acoustic ceiling tiles and printer covers are recommended for noise control but carpeting is not because it can cause a static electricity problem. Fire protection should be the chlorinated halogen type to avoid equipment damage.

Air conditioning need not maintain the meat locker environment required for older machines, and most equipment will operate up to 80°F with no problems. Air conditioning requirements are 3500 Btu/hr/kW of power consumption, in addition to any occupancy and exterior heat loads. The prime objective should be to maintain a constant temperature. Humidity should be maintained at 50-80% to control static electricity and possible condensation problems.

Room layout not only must consider operator's access to controls but also such mundane items as equipment repair access and floor cleaning equipment requirements. Raised flooring is the only practical way to provide mid-floor electrical outlets and interequipment cable runs without unsightly overhead wiring. The raised floor also permits a hidden ground grid and the flexibility of adding or rearranging equipment without electrical remodeling. Often, the control room is remote from the process and devoid of windows. However, the control room should be placed where the operator can view the process and have access to critical manual backup controls where possible.

Electrical

Electrical power system selection must be matched to the control system requirements. If all or part of the control system must operate at all times, an uninterruptible power system is required. These systems derive power from an inverter driven by a bank of batteries. During normal conditions, the battery charging system supplies enough power for both the inverter and whatever charging power might be required. These systems are rated in terms of the kilowatt-hours that can be supplied and recovery time to recharge the batteries. The UPS system can be designed with one unit for each computer or can be sized to supply only the critical process computer.

If it is permissible for a computer to cease operation during a power outage, power conditioning should be provided as a minimum. Special transformers that regulate voltage and eliminate electrical spikes are a must in the process environment. Electrical noise accounts for at least 90% of all unexplained system failures. Electrical noise can be introduced either through the power source, conductively through the grounding system or by radiation from sources of interference.

The computer room floor should have a ground grid that can be electrically connected to the "green wire" ground or to a separate earth ground. The grid not only provides a distributed electrical ground but also serves as a partial radiation shield. Grounding the grid at a single point minimizes the possibility of lower-frequency ground loop noise. Similarly, all shielded instrumentation cable should be grounded at one end only to avoid ground loops.

CONTROL SYSTEM HARDWARE

SUMMARY

From a hardware point of view, a computer control system consists of many and varied elements, depending on the specific needs of the application. In general, the system is composed of a mixture of the following:

- Computer
- Peripherals
- I/O equipment (local or remote)
- Telemetry
- Panels and terminations
- Instruments
- Control devices

The specific configuration, whether centralized, distributed or hybrid, will utilize various components as dictated by the application.

CHAPTER 4

CONTROL SYSTEM SOFTWARE

Terrance M. Brueck, PE

INTRODUCTION

Software is a highly important, but often overlooked, aspect of a successful computer control system. Many people fail to recognize or understand this, which can result in the downfall of the system. Too often, if the "system doesn't work" it is because the software has not been made to do what is expected.

If we question how something is accomplished in a computer control system, more often than not the answer will involve "software." As there is no "black box" to point to for explanation, we eventually find ourselves enveloped in a "software haze." This chapter will attempt to remove some of the mystery, dispel some of the myths and clarify the issues on software. In addition to defining process control software and its use, we will discuss how to procure it, control it and maintain it. A successful control system requires successful software.

Perspective on Software

When someone walks into a room and flips on the light switch, one expects the lights to turn "on." Although one cannot "see" the electric current flowing through the wires, one can see the wires (behind the wall). Control of the light is accomplished by hardware and the path of connection could be traced via the wires.

If the same control were accomplished by software, the path of con-

70 PROCESS CONTROL COMPUTER SYSTEMS

nection could not be traced via the wires. All wires eventually would go through a black box (i.e., a computer), and even tracing cables, printed circuit cards and computer chips would not reveal the path of connection. What would define this path of connection is a description of the program in that black box. Either a listing of the instructions or a flowchart of the logic would reveal the connection between the light switch and the light.

Figure 1 shows a flowchart of the light control. This depiction of the logic is very simplistic and only considers "high-level" decisions. No details regarding where the information comes from (inputs) or goes to (outputs) is presented. However, it does present the function of the light control.

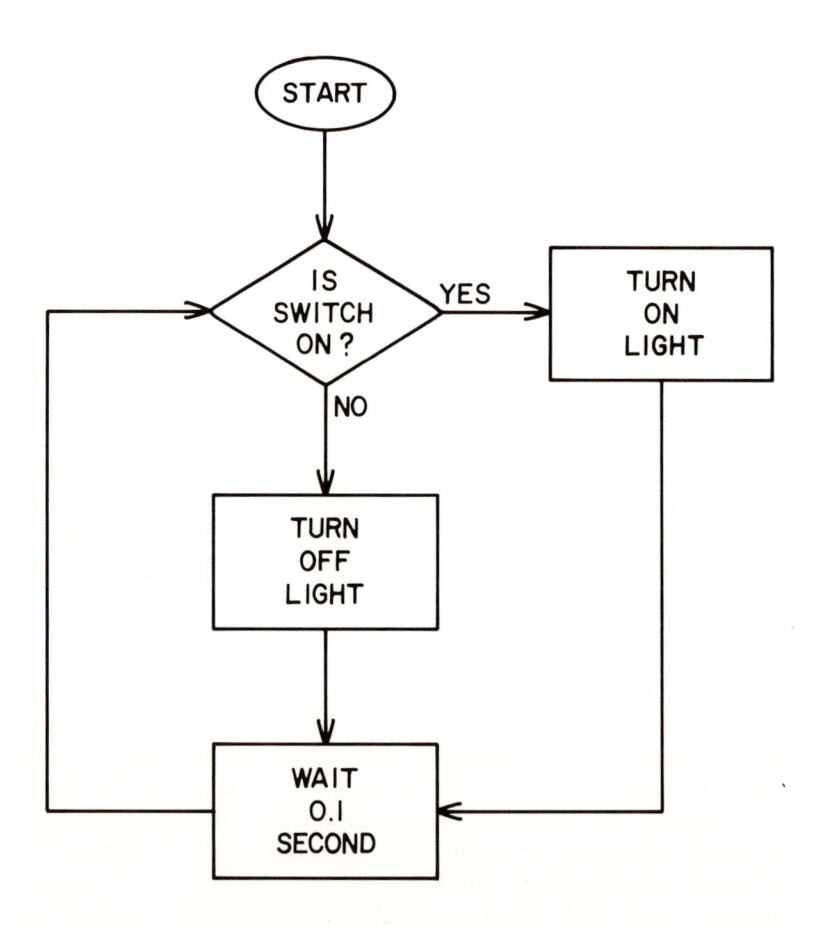

Figure 1. Flowchart of light control.

The same concept can be presented in a language-type format, such as:

>   DO ONCE PER 0.1 SECONDS;
>   IF SWITCH ON,
>   THEN TURN ON LIGHT.
>   ELSE TURN OFF LIGHT.
>   END

The control result should be similar to that of the flowchart.

Let us introduce several switches labeled A, B, C and several lights labeled X, Y, Z. If we want to "wire" a certain switch to a certain light we could define this via our language, such as:

>   DO ONCE PER 0.1 SECONDS;
>   IF SWITCH A ON,
>   THEN TURN ON LIGHT X.
>   ELSE TURN OFF LIGHT X.
>   END

To reconnect a different switch to a different light, or more than one light, would mean rewriting the program. This capability for change is one of the most powerful attributes of software but also one of the most difficult to manage effectively.

Today's computer control software will allow us to program the application in such a "high-level language" if the necessary software tools are in hand. These software tools have not always been part of a process control computer system, so it would be helpful to see how they have evolved. Then we can better understand the components of control system software and what is necessary or desirable for a given computer control application.

Control Software History

Since the late 1950s, when digital computers were first applied as "data logging" systems, process control (PC) software has existed in some form. These early systems, usually based on data processing-type computers, required the user to create his or her own real-time software to accommodate timing constraints and external hardware "interrupts." To meet desired response times, this was programmed in a low-level assembly language, which efficiently converts into machine code instructions. However, this required a significant amount of time from experienced programmers to develop even simple monitoring and control programs.

As real-time computer systems began to emerge, the computer manufacturers developed real-time operating systems, which allowed programmers to be less concerned with the management of the computer system resources. Even so, most programming was still performed in assembly language to allow programs to execute rapidly and utilize a minimum of the precious core memory. More complex mathematical computations were programmed using *FORTRAN,* when necessary.

With advances in computer hardware, speed and memory size became less of a constraint, and more application programs were written in compiled languages, usually *FORTRAN*. This still required personnel experienced in real-time programming to effectively develop control software.

The next advancement in control system software was the development of "high-level" languages. These process control languages allowed process engineers to do application programming with a minimum amount of training. These software packages have continued to expand in capability and become more "user-friendly." Applications for control strategies, graphic displays and reports now can be programmed in English language-type or fill-in-the-blanks languages, some of which are user-interactive through a terminal.

To summarize the development of process control software, Figure 2 presents several "generations" of the level at which a control application is implemented on the computer. As the capabilities of the software packages continue to expand, the computer itself becomes more "transparent" to the user. We can concentrate more on the process to be controlled rather than on the software to control it.

A dangerous temptation is to immediately begin programming because it is easily changed later. We must be cautious not to ignore proper analysis and design of the control application software and use the same care as if we were programming back in the first generation. Ease of programming is no substitute for proper control design.

CHARACTERISTICS OF CONTROL SYSTEM SOFTWARE

From the example of the light control program, several characteristics can be identified that make control system software unique and different from batch-type data processing software.

1. **Real-Time Response.** To the outside world (a person or process), it appears that the response is immediate or continuous and not a sequential step-by-step program. It appears that more than one program runs simultaneously.

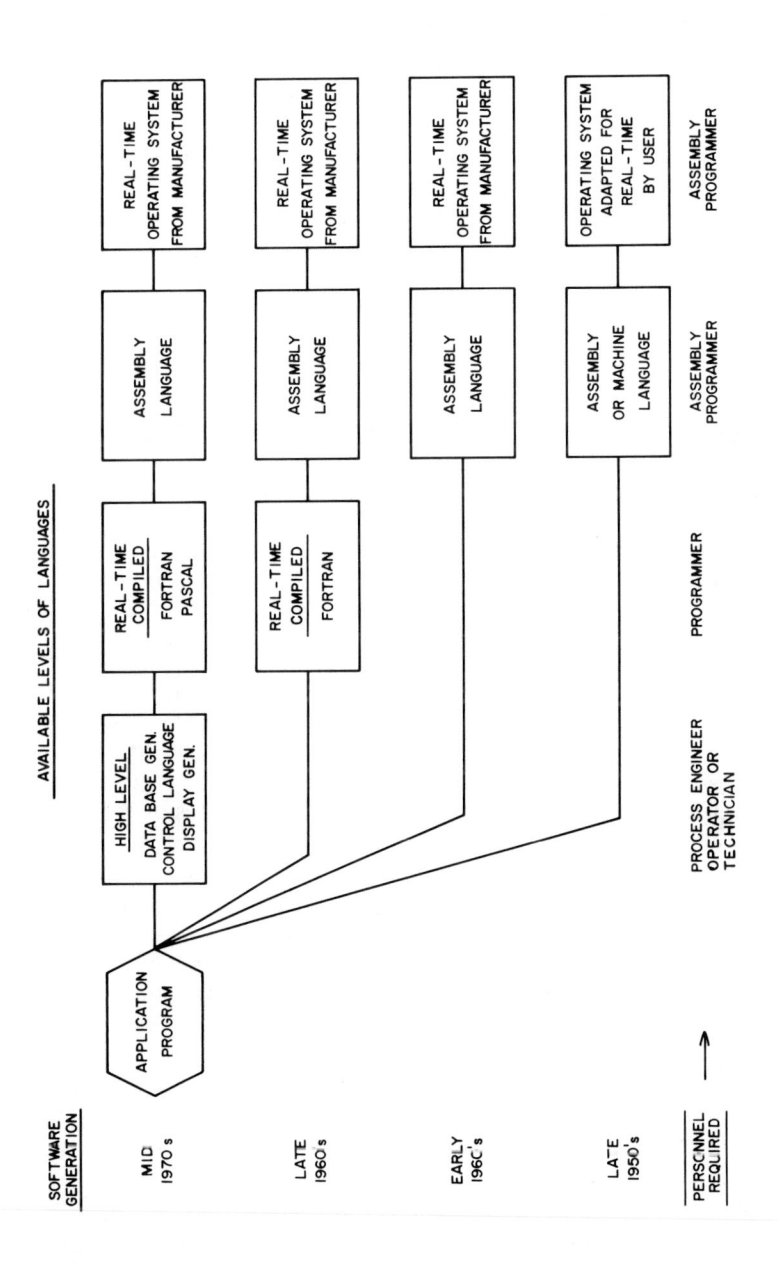

Figure 2. Generations of process control software.

2. **Interrupt Driven.** The program executes periodically when a scheduler (another program) directs it to run. This interrupt is time-based in software, although it could be a hardware interrupt, such as a console keyboard action. Both hardware and software interrupts are prioritized in order of importance to avoid conflict when two programs try to execute simultaneously.
3. **Repetitive.** It is not a one-time or "batch" program that terminates after executing once. It runs again like a continuous loop.
4. **Input/Output (I/O) Processing.** Although not apparent in our simplified example, other programs make data available to the light control program. These processors also run continuously to update I/O information to and from the external devices or sensors.

These characteristics of control system software can be understood more easily by looking at the components of the software. By defining the purpose and action of each component, one can see how it all works together. This will be done in terms of our "high-level language," but first the different types of software must be classified.

Types of Software

In our discussion of software history, we looked at different "generations" or levels of software capability. Each succeeding generation made it easier for us to program our application. Initially, the writing of application software required the same skill and expertise as the writing of system software. Later generations allowed application programming by someone with less programming experience. These types of software may be defined in simple terms as:

1. **System Software.** These are the software "tools" available for implementing the specific control application programs.
2. **Application Software.** This is the software "product" resulting from use of the system software for programming specific control strategies, video display terminal (VDT) displays, printed reports, etc.

By our definition of these types of software, we have implied that a "user" of the system software is necessary to develop the application software. Given a specific control application and the proper "tools," the user can develop the "product" as presented in Figure 3.

If the system software adequately meets our needs, then it can be used to generate the application software. If the system software falls short, two choices remain:

CONTROL APPLICATION

"TOOLS" → → "PRODUCT"

SYSTEM SOFTWARE   APPLICATION SOFTWARE

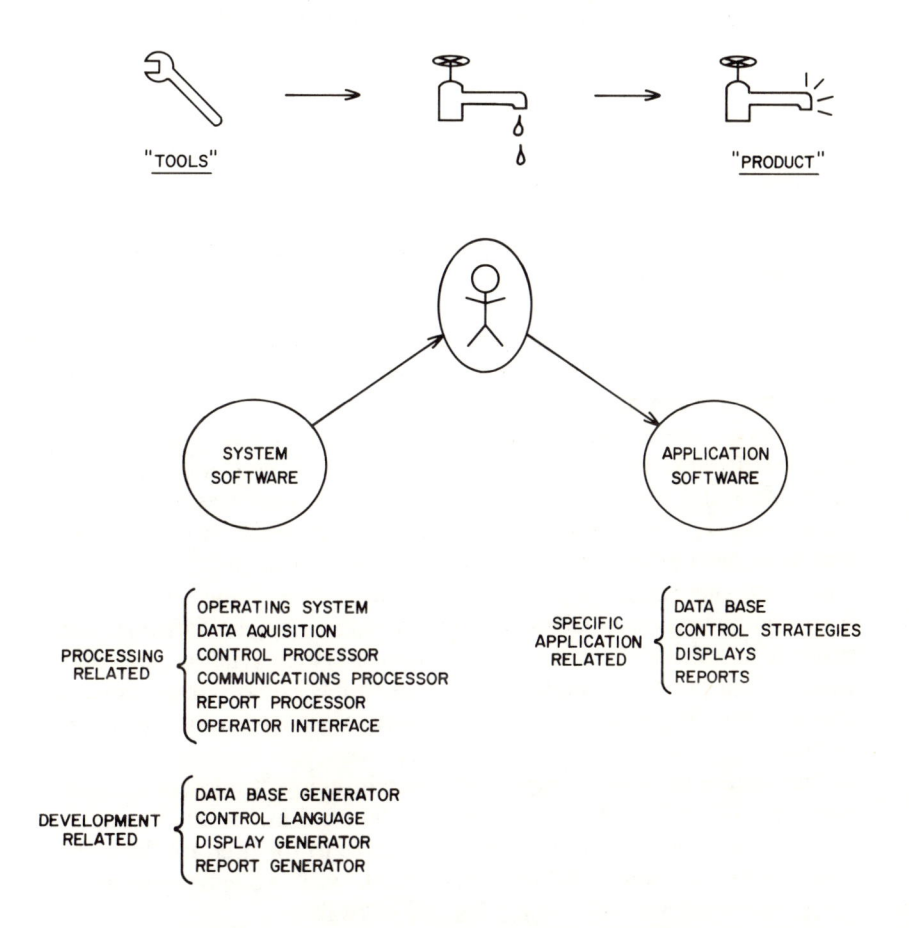

PROCESSING RELATED
- OPERATING SYSTEM
- DATA AQUISITION
- CONTROL PROCESSOR
- COMMUNICATIONS PROCESSOR
- REPORT PROCESSOR
- OPERATOR INTERFACE

SPECIFIC APPLICATION RELATED
- DATA BASE
- CONTROL STRATEGIES
- DISPLAYS
- REPORTS

DEVELOPMENT RELATED
- DATA BASE GENERATOR
- CONTROL LANGUAGE
- DISPLAY GENERATOR
- REPORT GENERATOR

Figure 3. Types of software: system vs application.

1. Develop the necessary "tools" by changing or adding to the system software to meet application requirements.
2. Develop unique software in lieu of system software that meets the application requirements but does not necessarily provide "tools" for other applications.

The early generations of control system programmers, being short of "tools," usually made the second choice to meet their application requirements. Those generations referred to that software as "application software," although it was mislabeled. Actually, it was a substitute for not yet developed system software. A further look into the system software and application software gives a better understanding.

System Software

As stated previously, if the system software "tools" are adequate, we can implement our application software without changing the system software. Further, the majority of this system software should be "transparent" to us, the users. We only need know our high-level language to program our application. Let us see what happens "behind the scenes."

The system software typically originates from two sources: (1) the computer manufacturer, and (2) the control system vendor. These can be the same, but usually the control system vendor purchases the computers from a computer manufacturer.

Normally, the computer manufacturer supplies only the operating system software, which is unique to the computer hardware, along with assemblers and compilers/interpreters for such languages as FORTRAN, BASIC, PASCAL, ADA or others. These are not the process control languages that have been talked about. Our requirements for application software should not necessitate use of these languages unless other system software components are inadequate. The remaining system software components usually are provided by the control system vendor. Sometimes independent "software houses" also will supply some system software components. Some large corporations have their own in-house control software group.

The system software components contain many "transparent tools," which are related to the processing that goes on continuously behind the scenes. The components that are more visible to the user are those that relate to development activities. These are the visible components we use when we create our specific application software.

Figure 4 exemplifies the system software components related to processing functions. These components can be identified as follows:

1. **Operating System.** This is the real-time monitor which, on a priority basis, schedules and allocates the computer system resources such as main memory, disks, tapes, printers, etc. All the necessary handlers, drivers and schedulers allow two or more operations to occur with apparent simultaneity.
2. **Data Acquisition.** This is input and output processing of data to field instruments and devices. This software converts signals to recognizable data values in engineering units (e.g., m^3/sec), checks for errors in signal ranges, checks alarm limits and can digitally filter the data to eliminate "noise."

CONTROL SYSTEM SOFTWARE 77

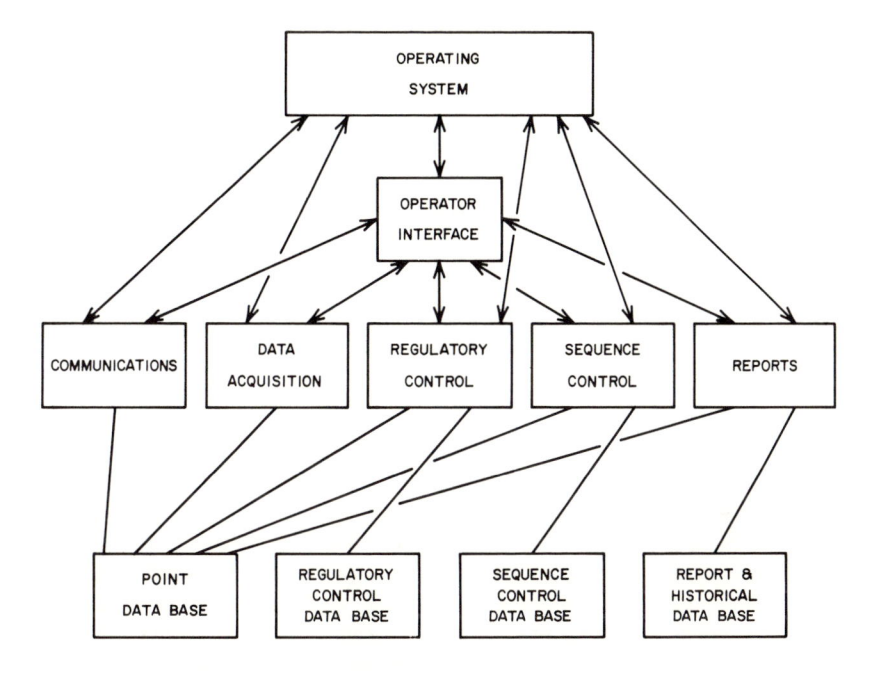

Figure 4. Example system software components.

3. **Regulatory Control Processor.** Algorithms emulate analog controllers in addition to arithmetic and logical operators. This software provides traditional proportional, integral and derivative (PID) control, as well as other control and calculations. Linkages to the data acquisition inputs and outputs and other system software provide necessary functions for tuning, setpointing, etc.
4. **Sequential Control Processor.** Also called batch control, this performs discrete actions in sequential order. Our light control example uses this system software in its decision-making functions.
5. **Report Processor.** This refers to the printing of information, such as alarm logging, or recording and retrieving of data for summary reports. Data stored over time by this software then can be summarized through calculations to print periodic averages, totals, minimums, maximums, etc.
6. **Communications Processor.** This is the link to other computers or terminals, such as input/output multiplexers. This software transfers data and does the "handshaking" between computers. This is especially important in distributed control systems for communication on a data highway or local area network.
7. **Operator Interface.** This provides the operator with a "window to the process" through the computer. This software allows operator keyboard actions to call up graphic or alphanumeric VDT displays, which

present current data with automatic refreshing of live data. A hierarchy of displays typically allows paging and easy selection of desired displays. Control is accomplished through the keyboard.

The system software components related to development activities are the tools we use to implement our specific application software. These are the "high-level" or "fill-in-the-blanks" languages we will use. In the light control example, these development tools were used to "wire-up" our controls. Once "wired," the processing system software made it work accordingly. The system software components related to development functions can be identified as follows:

1. **Data Base Generator.** This defines I/O process points or "non-wired" internal points. This software creates the point data base that is accessed during system software processing functions. The system software can create the data base in an interactive method with the user or in a non-interactive (batch) method through a compilation procedure.
2. **Control Language.** This defines a control strategy through the use of modules or "English-like" statements. This software creates the regulatory and sequential control data bases and links to the point data base. The regulatory and sequential aspects may be independent and separate, or combined into one "language." The method also may be interactive or compiled.
3. **Report Generator.** This defines format and content of printed reports. This software also must define the storage of historical data if periodic summary reports are to be produced. A report and historical data base are set up through an interactive or compilation procedure.

Other system software not directly related to processing or development work but that should be included with the system is utility software and diagnostic software. These programs are used in troubleshooting, debugging and maintaining the computer system. These components can be identified as follows:

1. **Utilities.** These are programs that aid in debugging and maintaining other software. Some useful utility programs typically include file editor, manager and dumps; patch programs; disk/tape copy and transfer programs.
2. **Diagnostics.** These are programs to aid in maintaining or troubleshooting the hardware and include both online and offline diagnostic programs. Programs that exercise and check for malfunctions of all peripheral devices and the computer central processing unit (CPU), memory and I/O equipment also should be included.

Application Software

In the light control example, we already have seen some simplistic application software. If the system software "tools" are available, the application software for a more complex control problem also can be programmed using a "high-level" language. The software for a specific control application no longer requires the level of programming expertise it once did.

However, we must not let ease of application programming lead to lack of control definition and design. We must use a top-down approach in developing application software to ensure that the software will function as an integrated system and meet control system objectives. We will discuss some methods for controlling software development later in the chapter. First, we want to cover some aspects of control system software unique to particular kinds of computer systems.

Other Aspects

Until now, the issue of where the system software and application software reside has been ignored. We have discussed the software components as if they all existed in the same computer, which probably would be the case when dealing with a centralized minicomputer-based system. However, the advent of microprocessor and microcomputer-based remote units has led to distributed control systems. Another specialized case of these systems, which may be stand-alone or part of a distributed control system, is the programmable controller. Let us see how our previous discussion of control system software fits into each of these environments.

Distributed Control

The concept of distributed control is based on dividing the required control tasks among several machines while still providing a centralized point of operation. After microcomputers were first used as stand-alone controllers for small control applications, the potential for a network with centralized operational capability was realized. It was generally perceived that cost advantages may result from using several smaller computers rather than one large one. In addition, the "all the eggs in one basket" problems of a single computer system could be avoided.

The initial software capabilities of the first microcomputer control installations regressed to a first generation of software. Even given the sophistication of the "high-level" software now available, the microcomputer by itself did not have the capability and capacity (memory, speed, peripherals, etc.) to support it. Several generations of development systems were designed to allow software development work on a computer of greater capabilities, followed by implementation on the smaller computer system [1]. These methods were still a compromise from the system software capabilities discussed previously.

To provide the system software "tools" we have identified, the distributed control system must divide the software components between the central computer and the distributed computers. A typical split of the software components is identified in Table I. To allow the application software to be implemented on the distributed computers following development on the central computer, a computer-to-computer capability can be utilized. In this way, all the system software and application software previously identified can be implemented in a distributed control environment.

Table I. Division of Software Components

Software Components	Central Computer	Distributed Computers
Development Related		
System Software		
Data base generator	X	
Control language	X	
Display generator	X	
Report generator	X	
Application Software		
Data base		X
Control strategies		X
Displays	X	
Report	X	
Processing Related		
System Software		
Operating system	X	X
Data acquisition		X
Regulatory control		X
Sequential control		X
Report processor	X	
Communications processor	X	X
Operator interface	X	

Programmable Controllers

Programmable controllers first appeared in the controls market as replacements for relay panel logic. The enhancements made to them since that time now allow their application to more sophisticated control, including emulation of analog-type control loops. Many of the attributes of the system software previously discussed are now available on programmable controllers to some extent. In addition, they also can be implemented in a distributed environment.

However, they remain a specialized case of microcomputer control and, as such, are well suited only for specific applications. Ladder diagram logic is their forte. We may be disappointed if we try to apply them expecting the system software and application software capabilities discussed previously.

Achieving a Successful Software System

There are many ways to try to achieve a successful software system—some work, some do not. In attempting to identify the pitfalls and shortcomings of some, we will discuss (1) who can provide the software, (2) specifying and procuring the software, and (3) testing and accepting the software.

In a previous discussion of system software, the components and their general capabilities were identified. For a given computer control application, we must determine our needs in each of these areas before beginning the process of acquiring the software. One should be able to answer the following questions:

1. What are the control and monitoring requirements?
2. What capabilities for displays and reports are desired?
3. Who will maintain or change the software after it is initially implemented?

Different "high-level" languages obviously will have different capabilities. We need to evaluate these and other software capabilities for our particular application.

Software Acquisition

With the exception of the operating system software furnished by the computer manufacturer, there are two choices in acquiring the system software and application software: (1) procure it from the control system vendor, or (2) develop it in-house. There is a preferred method for both system software and application software.

System Software

The system software is closely tied to the hardware in areas of I/O interfacing to field signals, operator keyboards and VDTs, and data highway communications. The so-called "transportability" of this software from one computer to another is therefore lost. In addition, many person-years of effort have gone into the creation of this system software. For these reasons, the system software should be procured from the control system vendor.

Application Software

As discussed before, if the system software "tools" are provided, the application software can be programmed in a "high-level" language. The application is better understood by a process engineer or control engineer than by a control system vendor. If it is a yet-to-be-built facility, the control design engineer can define the control requirements. For these reasons, the application software can be programmed more effectively by the control engineer than by the control system vendor. Although it is possible to specify the application software to be provided by the vendor, it generally will be more efficient, less costly and provide better results through the engineers.

Specifying and Procuring

After the control system needs and requirements have been identified, two options are available for procuring the system software: (1) select and purchase, or (2) competitive bidding. The importance of procuring field-proven software cannot be stressed enough.

An evaluation of system software available from vendors should be performed in either method of procurement. Using a direct purchase method, the purchase price should be negotiated with the selected vendor and include his proposal as part of the contract.

For competitive bidding, a detailed specification for the system software reflecting the control system requirements should be prepared for bid documents. By using a two-step bid process and soliciting proposals, a vendor can take exception to a portion of the specification and bid his standard, field-proven software, rather than creating special new software. The two-step process then allows for evaluation of proposals and determination whether any exceptions taken will not meet the control needs. After the qualified bidder is selected, the specification can be revised to include any additional functions in the vendor's proposal and incorporate any exceptions to the original specification. The revised specification then should become part of the formal contract.

Testing and Accepting

Testing and acceptance of the system software should include (1) a factory test, (2) a field test, and (3) an availability demonstration period. Any corrections required to successfully meet the factory test procedure must be made prior to shipment. The field-supplied software then must match what was tested at the factory; the field test will confirm this. The availability demonstration period, usually lasting several months, is necessary to ensure that no "bugs" crop up that were undetected in previous tests. Final acceptance follows successful completion of the availability demonstration period.

To properly test the development "tools" of the system software, one must program some of the application software prior to the factory test. This will require the software training to be completed earlier in the project to allow implementing of data base, control strategies, displays and reports for the tests.

One should also allow sufficient time to complete all the necessary application software for control of the process before final installation.

SOFTWARE DEVELOPMENT AND MAINTENANCE

A procurement process has provided the necessary system software from a control system vendor. One must be concerned now with the development and maintenance of the application software, which includes controlling the software development, staffing requirements and procedures.

Controlling Software Development

Successful development of the application software must include the proper definition, design, programming, installation and testing phases.

The importance of a top-down approach becomes more significant with the size of the total control system. The control objectives should consider the overall picture starting at the top as:

- Plantwide
- Areawide
- Process
- Subprocess
- Equipment group
- Final element

One must be cautious not to construe ease of programming in a "high-level" language as a reason to waive proper control definition and design. Successful software requires the same steps no matter what language level is being programmed. A more complete discussion of the necessary steps for a software development plan is given by Metzger [2].

Staffing Requirements

Chapter 10 presents an overall staffing picture for the computer control system. The application software development phase for a large project will exceed the normal staffing level for software upgrading and maintenance. Control design engineers or process engineers familiar with the particular process and knowledgeable in computer control systems can provide the additional staffing during the development phase. This avoids the communication normally required between the control designer and the computer programmer, which is necessary if the control system vendor provides the application software. This is highly recommended due to the difficulty of specifying control software in sufficient detail to communicate the control designer's intent. Extensive process knowledge is necessary on the part of anyone programming application software if it is to result in a successful control strategy.

Procedures

Successful software will require specific procedures to be followed for effective management and documentation. Management includes both management of the people and management of the materials. People management is required for the software for three basic reasons:

1. Solicitation and tracking of input from system users regarding prob-

lems or desired improvements. These users can include operators, process engineers, laboratory personnel, or maintenance personnel.
2. Authorization and enforcement of methods or changes. This will help prevent maverick system users from "doing it their way."
3. Training following changes or retraining. System users must stay current with the way the application software works.

Management of the materials includes routine tasks, as well as update tasks when changes are made. Software-related materials applicable to these tasks that must be organized and updated can be outlined as follows:

1. Routine tasks
    Report verification routing and filing
    Tape and disk backup and filing
    Historical tape generation and filing
    Downtime schedule
    Log book
2. Update tasks
    Change request forms
    Notification of change forms
    Software documentation including:
        Listings
        Flowcharts
        Strategy diagrams
    Operation manual
    Tape and disk backup or source

These tasks must be carried out by designated personnel under the direction of the system manager (see Chapter 10). One must make a concerted effort to keep the documentation and backup up-to-date so that a failure does not become a disaster, especially when the development system software allows interactive changes to the data base, controls, displays or reports. Although a convenient feature, this allows changes to be made without any listing being generated or source materials being updated. Such sophisticated system software requires careful tracking of application software changes to ensure the current software status can be recreated.

SUMMARY

Through an example control application, we have discussed the flexibility and capability of control system software. This software is unique in capabilities for real-time responses and priority-based actions. With succeeding "generations" of software, one can program control applications using "high-level" languages. If the necessary system software "tools" are procured, the control application software can be programmed by process-oriented personnel with the proper training. Staff must be designated and procedures defined and carried out to maintain and continue to upgrade control software capabilities.

Do's

- Do initially identify control application needs for control system software.
- Do evaluate the "high-level" software available that can meet the needs.
- Do acquire the system software "tools" to allow application software programming by process-oriented personnel.
- Do procure field-proven "standard" system software from vendors whenever possible.
- Do define the software management procedures and carry them out for effective development and maintenance of control system software.
- Do identify and designate personnel responsible for control system software.
- Do realize that software is never fixed and final:
  Changes are always desired or required.
  Debugging is never complete.
  Easily made changes can be difficult to manage and control.
  Documentation must be kept up to date.

Don'ts

- Don't procure turnkey control system software that cannot be modified without vendor involvement.
- Don't let ease of application programming lead to lack of control definition and design.
- Don't expect software not to require attention by someone—it will!

REFERENCES

1. Toong, H. D. In: *Mini- and Microcomputer Control in Industrial Processes*, M. R. Skrokov, Ed. (New York: Van Nostrand Reinhold Company, 1980), pp. 71–77.
2. Metzger, P. W. *Managing A Programming Project* (Englewood Cliffs, NJ: Prentice-Hall, Inc., 1973).

CHAPTER 5

BENEFITS OF DIGITAL CONTROL

Alan W. Manning, PE

INTRODUCTION

This chapter presents the benefits of digital control, which are a direct result of the flexible nature of digital systems. The technological evolution of digital control is discussed and the prime forcing functions of the evolution are explained. The future direction of digital hardware and "user-friendly" software also is discussed.

Both tangible and intangible benefits are presented in this chapter. Due to the programmable nature of digital systems and the development of "user-friendly" software packages, these potential benefits are becoming easier to realize. The chapter includes a summary of these benefits, a formula for achieving them and a brief look to the future.

CURRENT TRENDS

Background

Process control automation has evolved along with digital technology over the last twenty years. In the early 1960s, analog control on a loop-by-loop basis was the accepted approach. Any optimization or decoupling of control loops was handled by the operator. Relays and timers were used for electrical interface. Successful installations were commonplace within the framework of the technology of the day.

Things began to change as plants became larger and more complicated. Coordination and operator accountability became more difficult. Manual decoupling, optimization and/or coordination became impossible and centralization of information became a necessity to stay productive, efficient and profitable.

Analog solutions to centralization were attempted first. These were successful, and in some cases the approach is still used today. However, the problem was usually high cost, while operator accountability was not enhanced. Hybrid systems were tried involving a digital computer added to an analog system in an attempt to add operator accountability and data acquisition. This helped solve the operator accountability problem but further increased the costs.

Next came the first version of digital control. Usually, the information was all "hardwired" to a central control room where a computer would gather the data and output control signals. It was learned quickly that the installed cost of this approach was very high, particularly when considering startup time and initial reliability problems caused by the massive wiring efforts. This led quickly to the use of remote multiplexers or remote terminal units (RTU), which would collect data in the field from hundreds of wires and transmit the information over a single pair of wires to the computer. This immediately saved a great deal in wire costs and made centralization a reality for many processing operations.

Digital systems with remote multiplexing solved the money and accountability problem, but a new concern arose as a result of previous experience—reliability. These multiplexers could not function without the master computer and, therefore, when the main computer or computers failed the whole control system was inoperative.

This reality and the advent of microtechnology quickly led to the development of the "smart" or "intelligent" RTU which can function in the field without the master computer. This means that reliability is improved while all the other advantages of cost-effectiveness and operator accountability are sustained. When the master computer is available, full centralization is achieved. When the master is not available, process control is maintained by the RTU.

Table I presents a subjective comparison of the various centralization control alternatives versus some important criteria. The table is presented as a guide only, and all comparisons are versus manual control. Because of the decreasing costs of hardware and recent "user-friendly" evolutions in software, distributed digital systems are becoming more and more popular.

The Future

As microtechnology continues to concentrate electronics into smaller and smaller packages at reduced costs, microcomputers will be distributed into the field to further decrease wiring costs and to increase reliability. It is unlikely that new technical advances will replace the digital method; however, packaging and software enhancements will make the computer more and more transparent. In the future, every sensor and control device may have its own microcomputer, which communicates with process equipment microcomputers, which communicate with control microcomputers, and so on. The hardware costs will continue to fall, the wiring savings will increase continually, the reliability will be ever improving and the results will speak for themselves.

POTENTIAL TANGIBLE BENEFITS OF DIGITAL CONTROL

The potential benefits of digital control are discussed below. These tangible advantages can be attained if the equipment is properly designed, applied, maintained and, most importantly, managed.

It must be stressed, however, that digital systems bring with them some new management challenges. They must be applied diligently. They do not work on their own and their inherent advantages are not automatic. They can be programmed to do what is needed. The questions often arise: "Who knows what is needed?" and "Who is in charge to make it happen?"

The tangible benefits of digital systems to be discussed are as follows:

- Capital cost avoidance
- Control wiring savings
- Labor cost avoidance
- Chemical savings
- Energy savings
- Centralization, yielding, production consistency

Capital Cost Avoidance

Cost/benefit analysis of automation and computer control systems historically has been based on anticipated savings in labor, raw materials, energy and improvements in quality control. This means of benefit calculation usually is adequate to justify a control system for a new pro-

Table I. Control Comparison Summary

		Analog			Digital		
Consideration	Manual	Local	Central	Distributed	Central	Distributed	Hybrid
Costs							
Equipment	Low	Moderate	Moderate	Moderate	High	High	High
Instruments	Low	Moderate	Moderate	Moderate	High	High	Moderate/high
Installation	Low	Moderate	Moderate	Moderate/high	Moderate/high	High	High
Operating	High	High	Low	Moderate	Low	Low/moderate	High
Maintenance	Low	High	High	High	Moderate	Moderate	High
Software	None	None	None	None	Moderate	High	Moderate
Expansion	Low	High	High	High	Moderate	Moderate	High
Savings							
Labor	None	Some	Some	Some	High	High	Some
Chemical	None	Some	Some	Some	Moderate	Moderate	Some
Energy	None	None	None	None	Moderate	Moderate	Some
Equipment Life	None	None	None	None	Some	Some	Some
Reliability	High	Moderate	Moderate	Moderate	Low/moderate	Moderate	Moderate
Flexibility	Complete	Low	Low	Low	High	High	Low
Optimization	None	None	None	None	High	High	Some
Space Requirements	Low	Moderate	Moderate	Moderate	High	High	High
Operator Qualifications	Low	Moderate	Moderate	Moderate/high	Moderate/high	Moderate/high	Moderate/high

Operator Acceptance	Good	Fair/good	Fair/good	Fair/good	Good	Good	Good
Management Information	None	None	None	None	Complete	Complete	Moderate
Methods of Control	Panels, crisis control	Panels, crisis control	Panels, crisis control	Panels, crisis control	VDTs, anticipatory control	VDTs, anticipatory control	Panels/VDTs, crisis control
Automatic Control Available	None, human intuition	Feedforward, feedback	Feedforward, feedback, cascade	Feedforward, feedback, cascade	Feedforward, feedback, cascade, adaptive	Feedforward, feedback, cascade, adaptive	Feedforward, feedback, cascade, adaptive

cessing plant or to upgrade a process presently using a minimal amount of instrumentation.

In the more advanced industries such as electric utilities, where maintaining required levels of productivity is mandatory, savings due to automation are now being examined in the area of reduction of capital investment because of tighter control [1]. Many companies that are using the more conventional approach to benefit calculation may be overlooking the major savings that can be attributed to deferment or a reduction in capital expenses.

Typically, most processes are designed conservatively with ample spare capacity based on historical load variations or expected peaking factors. By optimizing the performance of a process and reducing process variations, the required safety margin can be reduced, thereby yielding a savings in capital investment on new process construction. If an existing process is at or near maximum capacity, the process improvement provided by tighter control may increase the process capacity, which also can defer plant expansion and the associated capital expenses.

Numerous examples of improvements in process performance and increase in plant capacity through the use of computer control are given throughout the literature. Benefits have been shown in the power, chemical, petroleum, paper, cement and wastewater industries [2]. In an experiment at a wastewater treatment plant [3], computer control of an activated sludge process was compared to manual control on parallel process streams. In addition to the considerable energy savings, the process under computer control required one-half the capacity in the final sedimentation tanks to achieve the same effluent quality, which illustrates the significant potential for capital savings through improved process control.

Control Wiring Savings

The basis of the control wiring comparison described in the following discussion is work that was done for the U.S. Environmental Protection Agency (EPA) and published in 1980. The cost of control wiring for centralization was analyzed in that report for three alternative control systems and the five physical plant areas. The analysis starts with the calculation of the cost per pair foot of analog and discrete wiring.

Wiring cost calculations are derived from *Building Cost Construction Data 1977* [4]. The costs per pair foot were derived by summing material, installation, and overhead and profit (O&P) costs. Termination costs are $1.76 per pair.

Discrete Signal Wiring

Typically, this is wiring from output relays to field control relays, magnetic starters and solenoids, or wiring from status and alarm contacts from field equipment. Two #12 AWG stranded conductors constitute one pair. Seven pair are installed per 1.5-inch rigid conduit. Installed costs are $0.8753 per pair foot.

Analog Signal Wiring

This wiring is typically for 4–20 mA, or 1–5 V signals. Two #14 AWG stranded conductors with an overall plastic jacket constitutes one pair. Seven pair are installed per 1.5-inch rigid conduit. Installed costs are $0.9643 per pair foot.

Wiring requirements were developed from schedules of equipment mounted in the central monitor panel or the central control room for conventional and central analog control systems, respectively. The wiring for field-mounted equipment, which is independent of the control configuration (level transmitters, flow transmitters, etc.), is not included in the wire count. The wiring analysis for a digital control system requires a schedule of multiplexer locations to determine the pair requirements.

The wiring distances were developed based on varying plant areas. The process centers maintain the same relative proportions on a varying plant site area. The area of the plant site usually varies in a relationship to production. The land area (exclusive of expansion space) per unit of production was estimated from several actual plants.

Estimates of wiring costs for the alternate control systems are summarized in Table II for comparison purposes. Wiring costs for conventional analog control assumes some (limited) information is brought back to a central location. Wiring costs for central analog control are high because more information and control signal wires are routed to the central location. Table II shows substantial savings in wiring costs in digital systems because wiring is reduced by the use of remote (field) multiplexers. These savings, while partially reduced by the cost of the multiplexers, nevertheless remain substantial.

Operational Labor Avoidance

The primary objective of process centralization and automation is to reduce dependence on operational labor. Decreased operational dependence can be attained by adopting an "operation by exception" philosophy.

Table II. Wiring Capital Cost Summary ($)

	\multicolumn{5}{c}{Plant Area (acres)}
	1	10	50	100	500
Conventional Wiring[a]	10,021	24,320	69,606	164,058	574,749
Central Analog Wiring[a]	52,041	105,446	317,267	739,211	2,202,833
Central Digital Wiring[a]	8,565	14,349	28,900	72,662	150,815

[a]Does not include wiring between field equipment and local control panels or multiplexers.

Events that are out of the ordinary are detected by the control system and annunciated to an operator for action, thus reducing or eliminating labor required for routine monitoring of equipment and manual data logging.

The potential annual cost savings due to centralization are significant because the cost of labor for operations and maintenance (O&M) of a typical process facility can approach one-half the total operating budget. If a capital investment for centralized control is annualized over 20 years at 7% interest and compared to the annual salary of an operator ($20,000 per year), it can be shown that more than $200,000 in capital costs can be amortized.

To demonstrate potential operational labor reductions that may be achieved by centralization, the tasks associated with three actual operator positions were evaluated. Tables III–V show tasks for a processing plant and were obtained during field visits [5]. The approximate completion times for each task are estimated for centralized control based on the fact that the control system frees the operator from much of the routine equipment monitoring and manual data logging that otherwise would be required. The time estimates also were gathered from the field interviews at plant facilities.

Table III indicates that centralization can reduce by about 40% the time required for typical operational functions performed by process operators. Table IV indicates that typical process-attendant control functions and operating time can be reduced by approximately 41%. Table V shows that the operational tasks for a typical operations apprentice will not be reduced or affected by centralization. Checking of devices (preventive maintenance), sampling, testing and cleaning all continue to be necessary. Centralization saves operational man-hours predominantly

BENEFITS OF DIGITAL CONTROL 95

Table III. Process Operator Tasks

| | Estimated Time to Complete (min/shift) ||
Task	Conventional Control	Centralized Control
Check Compressor Equipment (motor, bearing, vibration)	30	0
Foam Check	30	30
Motor Drives and Other Equipment Checks	30	0
Pump and Motor Check (bearing, temperature, vibration)	60	0
Valve, Alarm and Status Change Logging	60	30
Control Loop Check #1 (four times per shift)	60	60
Control Loop Check #2 (four times per shift)	60	60
Air Flow Control Loop Check and Record	60	30
Control Loop Check and Record #3	60	30
Miscellaneous	30	30
Total Time	480	270

Table IV. Process Attendant Tasks

| | Estimated Time to Complete (min/shift) ||
Task	Conventional Control	Centralized Control
Process Equipment Operation	120	120
Outside Cleaning	30	30
Device Checks	60	0
Dissolved Oxygen Logging	30	0
Air Supply Logging	30	0
Flow Logging	45	0
Centrifuge Tests, Logging, etc.	60	60
Flow Checking and Logging	15	0
Miscellaneous (device checks, alarms, operator communications)	45	45
Total Time	435	255

in the process operator and process attendant categories. Operation by exception, made possible by the use of centralized control, reduces the total number of hours needed to make control decisions and log data. In the examples cited, the average person-hour savings can be as great as 27% in terms of actual operational functions and time.

96 PROCESS CONTROL COMPUTER SYSTEMS

Table V.   Tasks of an Operations Apprentice

Task	Estimated Time to Complete (min/shift)	
	Conventional Control	Centralized Control
Field Tests	120	120
Centrifuge Tests	60	60
Sample Line Flushing	60	60
Sink Cleaning	30	30
Probe Flushing	60	60
Composite Sample Pickup and Distribution of Samples	30	30
Pump Cleaning	60	60
Lamp Replacement (average)	30	30
Floor and Drain Cleaning	60	60
Total Time	510	510

Chemical Use Reduction

Reduction of the chemical dosages required for proper control has economic benefits in many processes. This conservation of chemicals can be optimized through automatic control systems that incorporate the following:

- Appropriate instrumentation
- Computer-based decision-making
- Effective control algorithms

One form of algorithm or control strategy practical to chemical control is load following. This involves the definition of the process in terms of a mathematical formula. This formula can be as simple as a proportional adjustment relative to the load (e.g., hydraulic load following) or as complicated as a logarithmic relationship describing a pH titration curve. Using the formula or model, the required chemical feed is calculated for actual process needs. These feedforward control models should be accompanied by feedback control to compensate for errors between the actual process load and the load predicted from the mathematical equation. This combination of feedforward and feedback control minimizes chemical costs and optimizes the process operation.

In Figure 1, the use of two types of load following is compared with constant rate chemical feed to show how load following can reduce the chemical usage. The concept of load following, when applied to oxygen

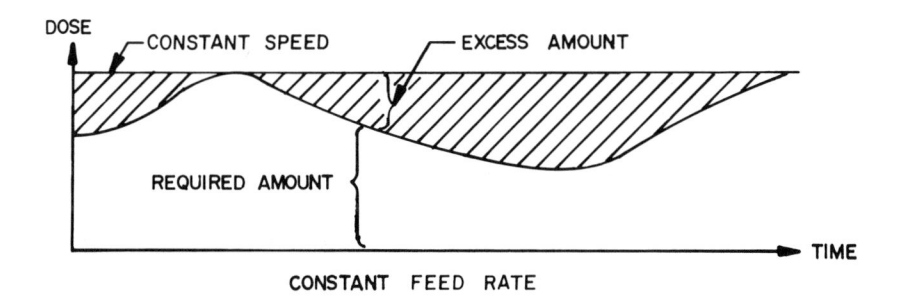

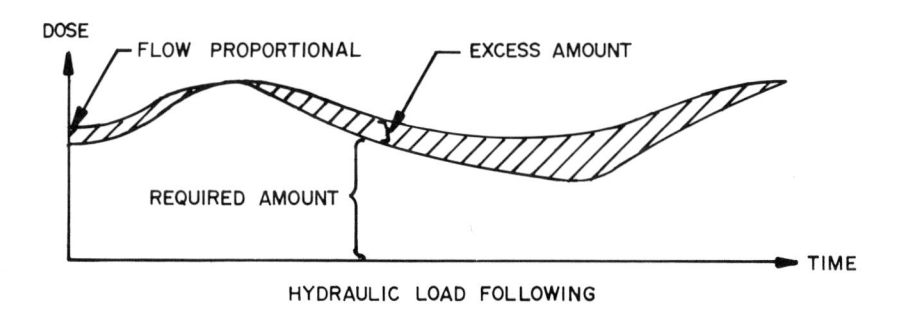

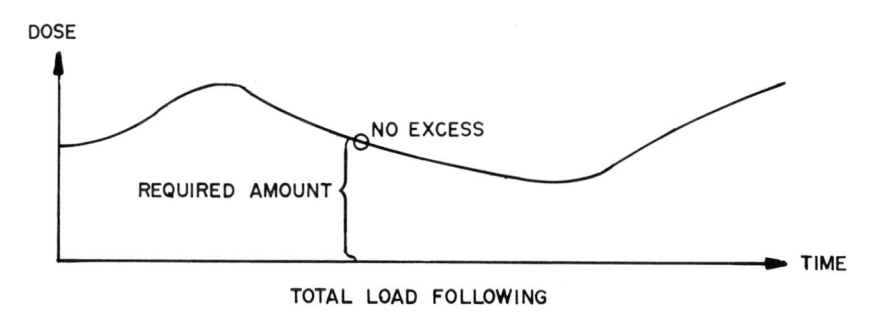

Figure 1. Chemical feed control alternatives.

control, saves energy by minimizing air flow in a manner similar to that shown for chemical feed. Feedforward control is based on mass flow of raw materials. The mass flow is translated to a required air flow, and the measured dissolved oxygen provides feedback trim. Dissolved oxygen control has been shown to yield a significant reduction in energy use when advanced control strategies are applied [6].

Conventional control systems typically control chemical feed on an open-loop basis. This means that an operator periodically sets a chemical

98 PROCESS CONTROL COMPUTER SYSTEMS

feed dose rate controller and that setting is maintained. Dissolved oxygen control also is conventionally implemented on an open-loop basis. The operator increases or decreases air flowrates based on visual observations or laboratory test results.

In estimating the average savings obtainable through the use of load following control, demonstrated experiences can be the only guide. In a study on dissolved oxygen control [6] involving 12 processing plants, an analysis of the improvement indicates an average power savings of 20%. This savings was realized using centrifugal compressors with adjustable inlet vane or suction throttling, as well as constant pressure control. The study analyzed multiple-plant sizes. As compressor power consumption can approach 50% of some plant power consumption, load following savings of 10% of total plant power consumption can be anticipated.

Load following can save a substantial amount of money when applied to chemical feed control [7-10]. To implement total load following in a practical way, a programmable system is most effective. Analog systems can provide some savings with single proportional control or feedback control. However, optimized control, yielding savings of 10% in chemical costs, can be achieved with digital systems.

Energy Use Savings

Power Demand Control Savings

Power company tariffs usually consider two factors in billing rates: (1) the amount of energy consumed, and (2) the maximum power demand. The basic charge is for energy used, whereas the power demand charges can be viewed as penalties or surcharges on the basic rate. The power companies monitor power demand during intervals of 15 or 30 minutes. The Cincinnati Gas and Electrical Company (CG&E) defines power demand as "...the kilowatts derived from the company's demand meter for the fifteen minute period of customer's greatest use during the month [11]." Power demand billing techniques vary throughout the country. CG&E bills on a specified percentage (50%) of the highest reading during a six-month period. Some other power companies bill the power demand charges on the highest demand reading during each month. There also may be a "summer ratchet," in which the demand reading during the winter months cannot be less than a specified percentage of the peak summer demand. Therefore, there are significant economic justifications for power demand control.

The total power load can be divided into two classes—base loads and selected loads. The base load includes lighting, small appliances and major loads, which are critical to system operation. Selected loads either can be reduced in power level or shed entirely for a short time as required to control demand peaks.

The basic concept of power demand control involves shedding a block of loads or preventing startups from occurring (so that demand will not exceed some upper limit) by deferring consumption of that block of energy to a period when the remainder of the electrical load is lower. In this way, peaks are lowered but energy consumption is not reduced. Figure 2 illustrates this concept. A 500-kw reduction in the power demand peak is obtained by deferring heavy energy use to a period of lower demand. This amount of power otherwise would be used to calculate a demand charge determined by the monthly demand peak multiplied by a rate that can be as high as $1.50/kw month. Therefore, keeping the power demand below the assigned peak demand can result in significant savings.

Power demand control cannot be implemented in a conventional or central analog control system because the complexity of the necessary control logic requires the capabilities of a digital computer. Industry has implemented power demand control techniques using computers. Applications have been reported [12,13] presenting successful installations where power costs were reduced by more than 10% by reducing demand charges. Numerous other industrial installations can be cited where savings like these have been demonstrated. It is estimated that a reduction of 5-10% of the average power cost can be achieved through the implementation of power demand control with digital systems. The potential savings in a particular plant will depend on the plant's operational flexibility (ability to defer or shed electrical loads such as blowers without significantly affecting overall plant performance).

Power Factor Control Savings

Line power consists of two components, active and reactive. The active power does the work; however, we also must pay for the reactive power because it increases the line current that must be generated and distributed by the power company, even though it does not contribute to the actual power used.

Reactive power results from the use of inductive and capacitive loads (i.e., motors). The relationship between active and reactive power is commonly referred to as the power factor. The power factor for a motor is a function of its load, as shown in Figure 3.

The penalty paid for a low factor is the "efficiency" of the energy

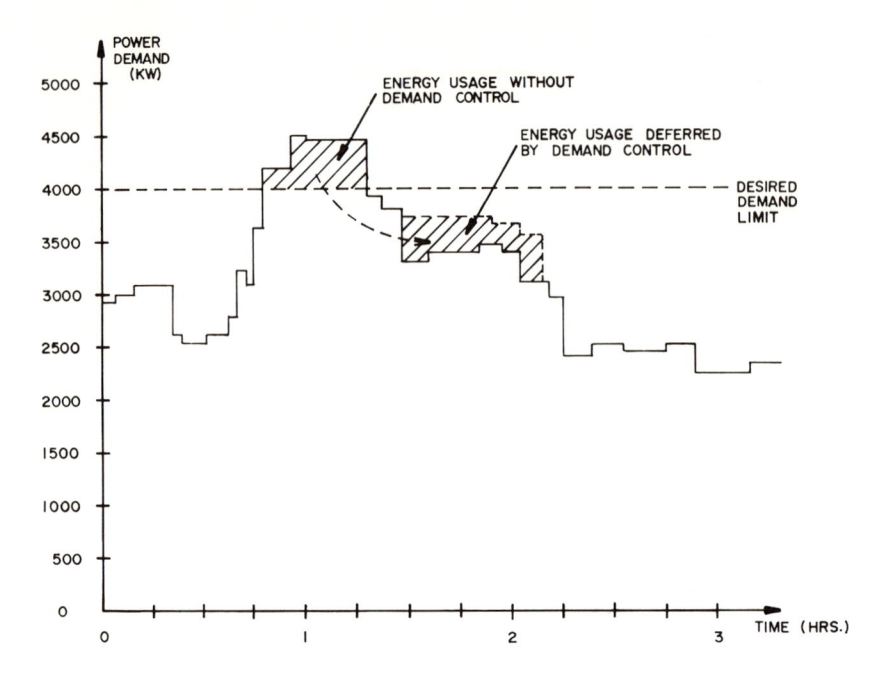

Figure 2. Power demand control.

purchased and, also, the cost penalty on the power bill. This typically occurs when the power factor is less than 0.85 or 0.90. This may amount to thousands of dollars a month for large industrial plants.

Fixed capacitors may be used to compensate for a specific electrical load. However, as the power factor is a dynamic variable that can vary continuously, it has proven worthwhile to provide continuous automatic compensation.

One processing plant saves approximately $100,000 per year based on dynamic compensation of power factor. A large pumping station with five 900-hp pumps was constructed for use during seasonal high-level flood conditions in the river that receives the discharge of a wastewater treatment plant. As this is an annual flood condition of short duration (and the flood condition may not occur every year), it was decided to use the motors for the dual function of pumping and plant power factor correction.

The motors can run as "synchronous capacitors" when not needed for pumping. For power factor correction, unloaded motors are operated and their excitation is varied to compensate for the plant power factor. The power factor measurement is made at the primary electrical sub-

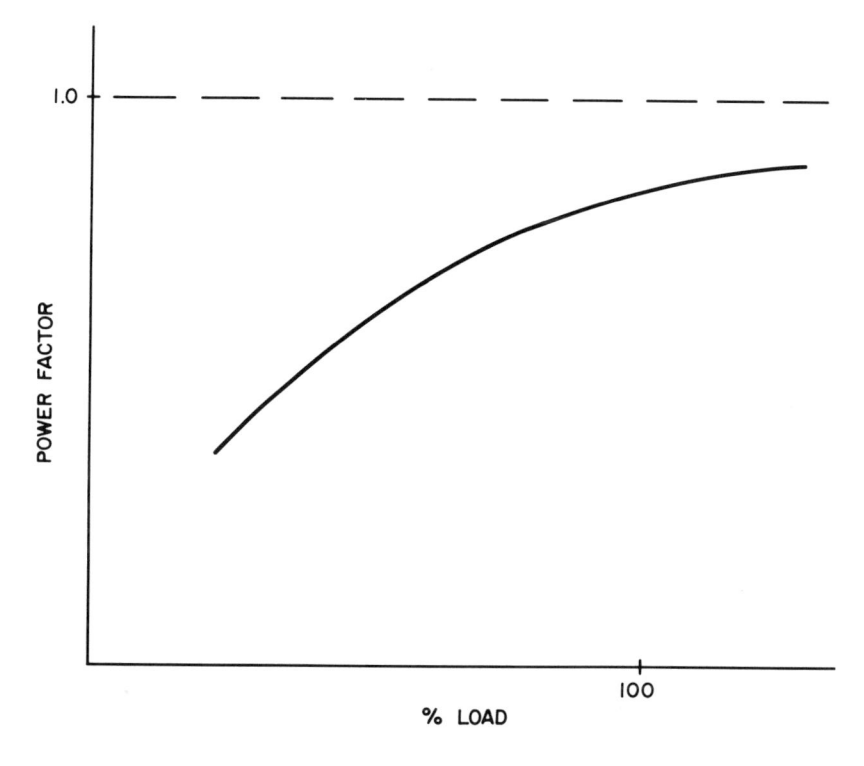

Figure 3. Typical power factor and motor load relationship.

station. This signal is received in the plant central control room, where the plant operator can select a setpoint and automatically maintain the power factor at a value between 0.7 and near unity.

Estimated savings on a percentage basis due to power factor correction is difficult to determine. In certain situations (as shown above), the savings can be substantial.

Centralization Yielding Production Consistency

Centralization, in a process control sense, means the availability of real-time process information at a single location to aid in process control decision-making. Although centralization aids in making consistent control decisions, the cost of signal wiring and conduit has been a major factor in prohibiting its use. With the advent of digital technology and the application of distributed digital techniques, centralization is more and more economically attractive.

Benefits of centralization are higher productivity, less off-spec product and greater process efficiency through savings in operating labor, chemicals and power costs. Since process control through centralization relies heavily on instrumentation, the potential labor savings initially is only moderate because the instrumentation requires higher maintenance labor intensity. The real labor savings is long term because a limit is placed on the operational labor costs.

As examples of the benefits to be derived, the cost of utilities was reduced by 3% for steam, 5% for energy (oil, gas, black liquor) and 8% for power in a paper mill due to centralization [14]. Crowell [15] cites three examples in which net annual savings varied from 24% to 42.5% of total power cost for a building, an auto plant and a steel plant. Centralization of a chemical plant utility operation is expected to produce savings yielding more than a 16% rate of return [16].

INTANGIBLE BENEFITS OF DIGITAL CONTROL

The programmable nature of digital systems offers advantages that can be discussed and qualified but not quantified. The advantages often are the true justification for obtaining a digital system. For example, in the electric utility industry it is not uncommon to justify the procurement of a $10 million digital control system based on the availability of information leading to increased security. It cannot be shown to save any money, but the system is required to manage the operation with the least possible risk.

Take a close look at these intangibles. They are very real, whereas the tangible benefits discussed before are all potential. They will depend more on specific instances and certain circumstances.

General Advantages

The general advantages of intangible benefits are presented in Table VI. The benefits are presented as advantages compared to analog or conventional control.

Modes of Control

With a computer system, multiple levels of control are available. In the first level of control, the computer, via programmed, changeable logic, can calculate the various setpoints, turn-on timers, etc. In the second mode, the operator will establish setpoints and turn-on modes and then utilize the computer to implement the decisions. In the third level of control, the operator will control devices directly through the video display terminal (VDT).

Reliability

Due to the various levels or modes of control, the computer system can be as reliable as an analog system. If an analog device fails within an analog loop, the analog loop is down. If a device fails in a computer loop, the operator can continue the control without the transmitter by using past or last-known values, or by using operator-entered values. If

Table VI.   General Advantages of Digital Control

Item	Traditional Analog Systems	Digital Systems
Modes of Control	Auto/manual	Computer directed Operator directed Manually directed Manual
Reliability	Good, loop by loop	Multiple levels of redundancy
Equipment life	No optimization	Reduced switching
Consistent Operation	Operator setpoints	Computer calculated and programmed
Use of Excess Capacity	None	Statistical programs FORTRAN
Flexibility	Limited	Programmable
Operator/Process Interface	Graphic panel Chart recorders	Color VDT, with graphics
Failure Considerations	Manual takeover	Enter bogey variable Hold on failure

the operator wishes, the computer can be overridden at any time. Standard provisions, such as the use of manual values if a transmitter fails, manual backup via the control console, and diagnostics that point out problem areas, complement the computer system.

Equipment Life

One objective of the computer system may be to eliminate on/off switching of high- and low-horsepower devices, thereby increasing equipment life. The computer system can be provided with the changeable programmed logic required to determine whether the device is needed at the particular time.

Consistent Plant Operation

When the computer programs are in control, inconsistencies in the operation are reduced. The control philosophy of the system does not change from shift to shift. The loops are tuned and these settings are noted in the record. Whenever an operator makes a change, that change is printed on the comment log for the manager to see. The computer is continuously using established relationships to calculate required chemical feed concentrations, plant flows, discharge pressure requirements, pump requirements, etc. When the computer encounters a problem, it requests operator assistance. Under normal conditions, the operator can spend time observing the operation of the system and doing laboratory analysis for entry through the VDT. This leads to consistent plant operation.

Use of the Computer for Other Purposes

With its computational capability, while the computer controls the process, statistical analyses and modeling can be performed as a low-priority task. For example, if the chemist requires statistical analysis for a particular date, he can use the "background" system to obtain the analysis.

Flexibility

The flexibility of the computer system is its greatest advantage. New process points can be added to the system without affecting the process control operation of the computer. Once added, they can be activated and added to the control system. They can be assigned to graphic screens online and used in control routines. New control models can be defined and activated online without affecting the operation of the computer system. If a new control mode is developed, it can be added without any hardware changes. This flexibility is provided because the system is controlled by software. If spare points are available to wire to the interface logic, only the wire has to be installed and terminated and the points specified in software.

Operator/Process/Computer Interface

The question of how the operator will control the system is important. The best method is via a color VDT. The color would indicate whether a device is running, in-alarm, out-of-service, not available, etc. The color of the flowline indicates function or process. The numbers for analog variables appear in real-time. Alarms may appear in red and flash until they are acknowledged, at which time they appear as a constant red. This method of communicating with the operator is highly efficient. The operator can look at the graphic screen and observe any point in any area of the plant. Such screens can be built online with a page builder program. A number of displays can be built and shown within any area of the plant.

Fail-Safe Mode

A properly designed computer system will provide for protection in instances of analog sensor failure. For example, a flow reading goes to zero, the valve controller "holds," and the software program goes into an alarm condition. This alarm is printed in the control room and simultaneously flashes in red on the VDT. The operator then can take action. However, the process is not disrupted due to the sensor failure.

Operational Data Reporting Advantages

The second category of intangible benefits of a digital system is the reporting, accountability and management information capabilities provided. This information aids in decision-making. This advantage affects not only management but operation as well. It could be called "management and operational information" because it entails noting when a pump starts or fails, when a device is out of service, running times of various devices or when an analog sensor fails, etc. This is typical management-type information and is not provided with typical analog instrumentation. These benefits are summarized in Table VII and then discussed individually.

Operational Changes

The computer monitors the running status of all devices to which it is wired throughout the plant. In addition, all operator actions are monitored and reported. Alarms and acknowledgments are recorded.

Analog Limit Checks

Sensed variables (flow, differential head, etc.) have a number of associated subchecks. Each has high/low operating limits, high/low alarm limits and high/low bad-value limits. If a sensor fails, the computer flags it as a bad value. If a sensor goes into alarm it becomes an alarmed value. If there is an operating limit violation, the sensor has a message printed that the operator must acknowledge.

Table VII.   Operational Reporting Comparison

Item	Traditional Analog Systems	Digital Systems
Operational Changes	No record	All status changes noted
Analog Limit Checks	Option	Three separate checks for operability
Alarm Reporting	Annunciator	Acknowledge required
Alarm Summary	No repeats	Chronological order
Lab Data Entry	Not available	For use in reports and operation
Report Generation	Not available	Change, add reports
Operating Reports	Not available	Almost infinite variety
Equipment Out of Service	Not available	Manual or out-of-service mode
Maintenance Report	Not available	Four tasks per device—acknowledge required
Trend Report	Not available	Variable points-variable frequency
Historic Data Storage	Not available	For retrieval later by customer

BENEFITS OF DIGITAL CONTROL   107

Alarm Reporting

If an alarm is generated in the system, whether it be an analog point, a pump failure, a control device that is not working, a sample pump that does not start or a level that goes beyond its operating range, an audible alarm is sounded in the control room. The device appears in flashing red on the graphic screen and in white on the alarm summary page. The operator can silence the sound with an alarm-acknowledge button. The specific alarm will have been printed out on the alarm printer. When the operator presses alarm-acknowledge, the device stops flashing on the screen and the acknowledgement and time are printed. When the device is returned to "normal" or "operational," the notation disappears and the flashing red signal returns to its "normal" color.

Alarm Summary

Alarm summary displays can be called that list the system's current alarms in chronological order. In addition to showing the alarm and a description of the alarm point, the time at which the point went into alarm can be displayed.

Laboratory Data Entry

A method can be provided so that hourly, daily and monthly information originating in a laboratory can be entered into the system. These data can have an associated date so that analysis that requires a long period of time can be entered for preceding days. There also can be a program for the daily entry of laboratory data. The information can be listed so that the operator is warned of the need to enter missing data. The laboratory data can be used in the control routines to optimize plant operation. Consequently, the operator can spend more time on analysis.

Report Generation

A report generator can be provided that will permit the operator to change operating reports. These reporting programs can enable the operator to modify and/or build new reports without the use of complicated assembly language or *FORTRAN* programs.

Operating Reports

The manner and format of reports that can be generated by a digital computer are many and varied. Suffice it to say that reports are possible that can meet any need. A typical list of reports for a given plant might include the following:

- Instant Plant Summary — snapshot of current situation
- Shift Report — summary of previous 8–16 hours
- Daily Report — summary of previous day's operation
- Monthly Report — summary of the month's important operating parameters, productions, efficiencies, etc.
- Annual Report — summary of a whole year's operation

Equipment Out-of-Service

A special display can be provided to illustrate all equipment that is in the manual or out-of-service mode. The display shall only illustrate devices that have been taken out of service via the computer/manual switch in the field or have been marked down by maintenance or by the operator.

Maintenance Reporting

The computer can monitor the running times of all directly wired devices and keep track of devices not directly monitored based on calendar time. Devices that do not have running times per se also can be monitored on the basis of calendar time. A maintenance report can be printed every morning. It would list the devices and specific maintenance tasks that are due to be performed at that time.

Trend Report

The operator can assign certain variables in the plant, along with a variable trend frequency of seconds to minutes, and the computer will print or display these points at that frequency for whatever duration is necessary. This report form is most useful for tuning control loops and optimizing the operation of the system.

Historic Data Storage

Data can be stored as averages for analog points as a permanent data file on magnetic tape or disk. The data will be stored so that it can be retrieved easily by an application program to be written at a later date for statistical analysis, trending or printing.

SUMMARY

Benefits

Automation and centralization can achieve many desirable objectives, the most important of which is a shift in labor from operation to ongoing maintenance. As operational labor is reduced, more effort can be concentrated on maintenance. The establishment of comprehensive preventive maintenance programs will greatly aid in achieving continuous, effective use of instrumentation and control equipment and assure more consistent production. Considering all of the tangible benefits described for centralization, Figure 4 shows graphically that digital systems theoretically can save money [5]. The larger the plant area, the more money that is saved due to the digital system. The important point is that digital systems offer the potential to minimize the rising costs of production.

Achieving Benefits

Experience has shown that all the benefits discussed earlier are possible, yet many digital systems fail to achieve their expectations. What must be done to realize these advantages? What is the secret of success?

Certainly the technical problems must be addressed. Assuming these are under control, however, experience has shown that failure can be avoided if a nontechnical formula is followed to the letter.

Component number 1 of the success formula is management commitment. It must be clear that management has selected the chosen system and is committed to its proper and total implementation. The direction is clear that the system has a purpose and an expectation.

Component number 2 of the success formula is staff focus. The system must have a "champion," who is responsible to make the system available for users (operations, etc.) and to assure that the system is applied properly. The key is the focus. The position can be either a line or a staff function, but the assignment and responsibility are to keep the "system" available and not allow changes outside of the prescribed system purpose. The focus is "system" and not "process."

Component number 3 of the success formula is a set of procedures that clearly identify the rules for the system. Who has the responsibility to perform what function? Who has what levels of authority? This is the rule book for the system manager and the users to follow, assuring system integrity.

110   PROCESS CONTROL COMPUTER SYSTEMS

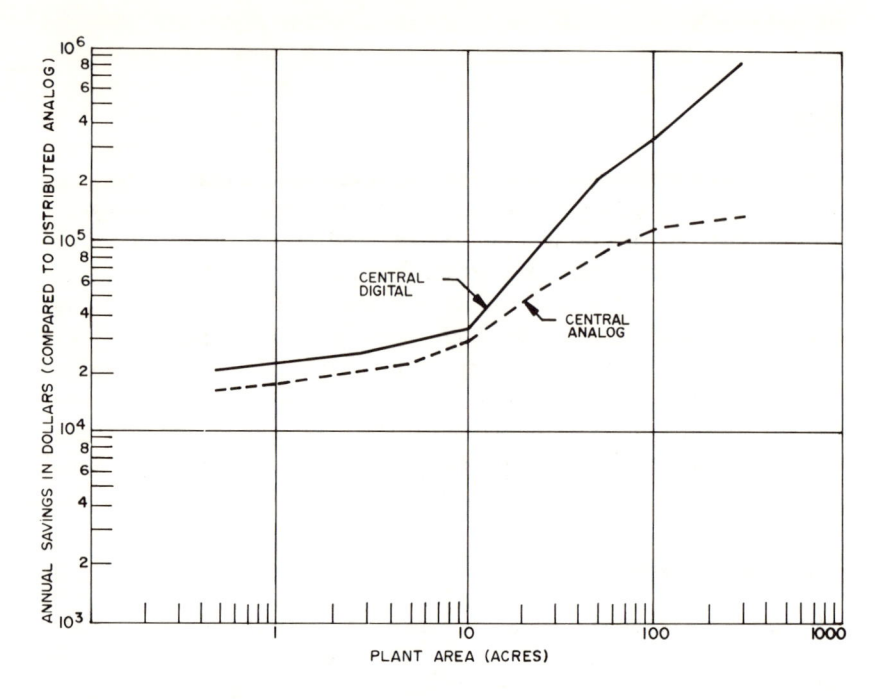

Figure 4. Estimated potential annual savings due to centralization — digital and central analog vs conventional analog.

These three components, taken in equal portions and gently mixed together over time, spell success:

(Management   +   Staff   +   Procedures)   +   *Time*   =   Success
(Commitment   Focus   )

This formula has been used consciously or subconsciously on every successful digital system project.

Future Possibilities

The 1970s have taken us from centralized hardware control to more and more distributed hardware configurations, and this trend shows no signs of abating. The future will find the further distribution of intelligence to lower processing levels, with microcomputers in individual instruments and control devices. The nature of field panels will change as all relay logic is replaced by small computers. It is likely that the time will come when the human being will never directly control a device without a microcomputer interpreting the manual commands and executing them. Designers of the future will find that computer control of a plant will include hundreds, or even thousands, of computers at various levels of the hierarchy, all necessary to achieve the objectives of cost reduction and accountability.

REFERENCES

1. Masiello, R. D. "Cost-Benefit Justification of an Energy Control Center," paper presented at IEEE Power Engineering Society Summer Meeting, Minneapolis, MN, July 13-18, 1980.
2. Stout, T. M. "Justify Process Control Computers: Selection and Costs," *Chem. Eng.* (September 1972).
3. Well, C. H. "Computer Control of Activated Sludge Processes (A Fully Nitrifying Process)," *Intech.* 26(4):32 (1979).
4. *Building Construction Cost Data 1977,* 35th annual edition of *Means Cost Index* (1977).
5. Manning, A. M., and D. D. Dobs. "Design Handbook for Automation of Activated Sludge Wastewater Treatment Plants," U.S. Environmental Protection Agency Report No. 600/8-80-028 U.S. Government Printing Office, Washington, DC (1980).
6. Flanagan, M. J., and B. D. Bracker. "Design Procedures for Dissolved Oxygen Control of Activated Sludge Processes," Environmental Protection Technology Series, EPA-600/2-77-032 (June 1977).
7. Molvar, A. E. "Selected Applications of Instrumentation and Automation in Wastewater Treatment Facilities," Environmental Protection Technology Series, EPA 600/2-76-276 (December 1976).
8. Molvar, A. E., Jr., J. F. Roesler, R. H. Wise and R. H. Babcock. "Instrumentation and Automation Experiences in Wastewater Treatment Facilities," Environmental Protection Technology Series, EPA-600/2-76-198 (October 1976).
9. Manning, A. W. "Direct Digital Control at the Iowa City Water Treatment Plant," *J. Am. Water Works Assoc.* 69(3) (1977).
10. Kron, C. "Direct Digital Control of the Iowa City Water Treatment Plant," paper presented at the National AWWA Conference, June, 1975.
11. Cincinnati Gas and Electric Company. Rate General Service, PUCO No. 16 (September 1976).
12. Thibodeau, G. J. "Peak Power Demand Savings Through Computer Control of Production Electric Furnaces," Digital Equipment Corporation, Maynard, MA (1977).
13. Crowell, W. H. "Power Demand Control for Lowering Operating Costs," in *Proc. ISA Annual Conf.,* Instrument Society of America (1974).
14. Scott, R. R., and R. Bradford. "Paper Mill Squeezes the Most Out of a

Complex Energy Mix," *Power* (April 1980).
15. Crowell, W. H. "Power Demand Control for Lowering Operating Costs," paper presented at the ISA Annual Conference, October, 1973.
16. Kaplan, J. H. "Computer Monitoring and Control of a Chemical Plant Utilities Operation," paper presented at the ISA Annual Conference, October, 1979.

CHAPTER 6

CAVEAT EMPTOR – LET THE BUYER BEWARE

Robert G. Skrentner, PE

INTRODUCTION

Several people who reviewed this chapter asked what "caveat emptor" meant. It was then that I realized that this warning had fallen into disuse. Because of the numerous consumer protection groups, both public and private, we have become somewhat complacent when purchasing products or services. We expect someone else to protect us from our own poor judgment. We expect that what we buy will serve its intended purpose. If it does not, we look to blame someone else for allowing a "shoddy" product on the market. We never seem to ask ourselves whether we investigated the product thoroughly or whether we applied it inappropriately. The explosion of product liability litigation reflects a similar attitude – someone else is to blame and I am going to punish them.

These changes in attitude have had significant ramifications for management personnel. It is still easy to make a mistake, but it is now more difficult to admit it.

Previous chapters have addressed the differences between process control and data processing, control system hardware and software, and the benefits of digital control. This chapter will look at some of the pitfalls, problems and causes of problems associated with digital control systems. By presenting a variety of case histories, I hope to alert you to some of the problems that may arise and their causes and thereby prevent their occurrence.

The reader is assumed to have a digital control system or be in the

114 PROCESS CONTROL COMPUTER SYSTEMS

process of acquiring one. The reader in the municipal marketplace may have heard of a number of unsuccessful digital control systems and probably is concerned that he avoids becoming another statistic [1-3]. For industrial managers, there does not seem to be as much reporting of failures unless they are spectacular [4]. However, secrets are better kept in the industrial arena.

Why do there seem to be so many problems associated with digital control systems? Why do some individuals seem to fear the mere mention of a digital control system? What experiences have caused this reaction?

Digital control components are extremely reliable. The space program has proven this. If solid-state devices were not reliable, we would not see the explosion in their use in radio and television, appliances, automobiles, etc. As the components of a digital control systems are reliable and as the control systems have been proven to work well in a number of applications, we must look elsewhere to find the cause of the unsuccessful systems.

SYSTEM ENGINEERING — WHO DOES IT?

For all types of control systems, the customer buys both products and services. The products consist of the control system components, including instruments, control devices, panels, and digital or analog control hardware. The services consist of the system engineering and programming to implement the desired control in the particular application and the services of people who will operate and maintain the components. These services may be rendered by the control system supplier, consultants, in-house staff or combinations thereof.

System Engineering — A Historical Perspective

For pneumatic and electronic analog-type control systems, much of the application engineering was performed by the vendor. This was normally in response to a need within a particular industry to control some portion of a process. In some cases, vendors acquired significant expertise in the application of their instruments and control devices to a particular process. It was very practical and cost-effective to rely on the vendors during the system design. Many of the control system designs came to rely on the "free" application engineering available from the instrument vendors.

In the early 1960s, when computers started becoming cost-effective for

SYSTEMS NOT MEETING EXPECTATIONS...

replacing conventional pneumatic and analog control, two parallel approaches to control evolved. Some control vendors went into the business of selling computers, and some computer vendors went into the business of selling controls. Users were sold the concept of digital control based on features other than control. This included replacing mapboards with video display terminal (VDT) displays, automatic generation of operating reports and a host of other features. Unfortunately, all parties learned quickly that the concept of free application engineering was no longer valid. The system engineering and programming costs equaled or exceeded the cost of the hardware. Some vendors went out of business, and many users ended up with far less than they had expected.

   The digital control systems available at the time consisted of control computers intended to replace the large central control panels. Control vendors approached the design of their central computers from the point of view of control loops. Small computer programs (modules) were interconnected to perform the desired control, much the same as the components of a control panel would have been interconnected. The computer vendors approached control more from the standpoint of data processing. In general, large computer programs were written to perform the required control. Some vendors attempted to combine the two approaches.

As a result, no two vendors approached control exactly the same way, which made the task of selection or specification of control systems very difficult. Dozens of vendors made relays and analog control panel components that performed almost identically. Designing panels was easy, as was bidding on panel manufacture and installation. For computer systems, the designer was forced to write a proprietary specification or a functional specification. He could not do the detailed design because each vendor approached control differently. In most cases, it was left to the vendor to perform the detailed application engineering. However, the vendor did not always know exactly what the designer had in mind for all the desired functions. Numerous arguments resulted along with the associated costs and delays in the project.

Vendors began to realize that the profitability of the company rested on the sales of equipment and not on engineering/programming services. These were necessary to get the sale. As a result, vendors attempted to develop erector set-type systems. The intent was to provide all the tools required for the user to do the application engineering while the vendor furnished only the computer hardware and support software.

Although a very reasonable approach, many users did not have the necessary personnel to apply the computers to the process, especially in the municipal marketplace. In addition, it was very difficult to estimate the size of the computer until much of the application work was completed. This was a "chicken or egg" situation for users. Users could not estimate the exact size of the equipment required until much of the application work was done; however, the application work could not be done until the computer vendor was known. Bids could not be taken on computer equipment until the size of the equipment was known.

Digital control technology continued to evolve, and in the mid-1970s vendors began to introduce distributed control systems. Two objectives were set. The first was to increase control reliability by distributing control around the plant. The second was to provide limited, definable control capability for each of the distributed controllers.

Based on their experiences and backgrounds, each vendor again approached the control differently. With central systems, most of the differences among vendors were in the software programs residing in the central computer. With distributed control, the number of combinations and permutations of hardware and software mixes to provide the distributed control was almost limitless. In addition, most users still wanted a central computer for overall plant monitoring and control coordination. How this computer was connected to perhaps a dozen other computers and how all the data were transferred around the plant and displayed to the operator were constrained only by the vendors' imaginations.

System Engineering Today

We still have not explained who does the system engineering required to implement a successful digital control system [5]. Vendors would prefer to sell hardware. They are not in the programming business by choice. Some vendors have used consulting engineers as representatives and for the applications work. However, either these representatives would prefer to concentrate on engineering service sales rather than vendor hardware sales or there is not sufficient staff to handle larger projects.

Most consulting engineers do not have the experienced programming personnel to perform the detailed application programming. Most system houses specializing in programming do not have the control engineering staff required to convert application requirements into software requirements. Most users do not have the staff to do the implementation and do not wish to acquire a large staff, which may have no more work at the end of the project.

By default, almost all municipal projects are implemented by the vendor, which is extremely costly from the owner's viewpoint. The owner must pay for engineering design for specifications, vendor time to translate the specifications into his system capabilities, engineering time to review vendor submittals to see whether they conform to the specifications and startup by the vendor and checkout by the engineer. This approach is at least twice as costly as other possible approaches.

Many private industries purchase the hardware based on an analysis of vendor control hardware and software capabilities as related to the specific application. Application programming and engineering are performed by in-house staff, especially when dealing with proprietary processes. Occasionally, vendor staff or other external assistance will be utilized to assist in implementation.

Most vendors are attempting to make the application engineering easier for the user by judicious design of the configuration capability of the digital control system. However, the vendor is still the best resource to provide implementation assistance.

The most cost-effective approach to digital control system implementation seems to be a consortium of the vendor, user and design engineer. In this arrangement, the designers and user are in the best position to know what the end result of the control system should be. The vendor is in the best position to assist the owner and designer in implementing the digital control system because he is most familiar with configuring and programming the components. The control system hardware may be acquired under competitive bidding based on functional specifications or preliminary planning studies. Once the hardware has been selected, the detailed design of the application programming can begin while equipment is being manufactured. The vendor, engineer and user can then configure the system and perform the application programming on a time and materials basis or a cost plus fee basis. In addition, the hardware procurement can limit the vendor to certain unit prices for additions should additional equipment become necessary as the detailed system engineering is performed.

A Crisis in the Making

The intent of this discussion of who should be responsible for the system (application) engineering for implementation of digital control systems was to create an awareness of a likely cause of many unsuccessful digital control systems. Although this may be a major cause of problems during implementation, the manager will rarely recognize it. What is seen is the end result of the problem. Typically, the first sign of a problem appears during vendor implementation. This may take the form of the vendor being behind schedule or requests for many changes and extras during the work. Often, the vendor has underestimated the scope of the work required [6].

While the whole topic of responsibility for system engineering may be a cause of many problems, a more serious cause is that the user was not sufficiently involved in the application engineering during the design and implementation of the project. As such, the user may have little ownership of design and little motivation to make the system work. The symptoms of this lack of involvement may become apparent when operators do not use the control system to its fullest potential; when computer control is abandoned in favor of control from local manual panels; when many complaints are heard that the system is not reliable or does not meet the expectations of the user; when the system is not maintained properly; or when management seems to lose interest in the success or failure of the system. By this time, there usually has been a large capital expenditure and it is too late to go back and start over.

How to Achieve the End Result

Five major activities lead to the end result. The control system must be planned, designed, implemented, staffed and managed. These activities may take place over several years and there are a number of areas in which problems can occur that could result in an unsuccessful system.

Each of these five areas will be addressed in the sections that follow.

PLANNING

Typical Approach

All planning results from a real or perceived need, which may take the form of improving existing product quality, producing more or new products or improving efficiency of operation. Once a need for planning is established, the planning talent must be acquired. This may be in-house staff or consultants. The planning process will set objectives, define and evaluate alternatives and recommend action.

Many factors can influence the planning process, such as preconceived ideas, external influences beyond the control of the planner, skills and abilities of the planner, and the time or budgetary constraints placed on the planner. These factors can be detrimental or beneficial to the planning process, depending on their ultimate influence on the decision-making process.

Planning Problem #1 – The Misdirected Plan

In this case, the owner thought he had a mechanical problem that was resulting in less than design performance of a process. He acquired the necessary mechanical engineering planning talent and proceeded to evaluate the process and test alternatives. Based on these tests, the process was modified at considerable expense. However, when placed in actual operation, the process performance did not improve as much as expected. The owner was expecting a 50% increase in performance – only a 20% increase was realized.

Further evaluation of the problem by a control systems specialist led to the conclusion that a major factor in the poor performance was an improperly tuned control system. Sudden changes in the flow to the process was causing the upsets. Minor modifications to the controls led to an additional 20% improvement in process performance.

Specialists from different backgrounds will approach a problem in different ways. In this case, the owner's preconceived idea of the cause of a problem led him to acquire planning talent, which tended to prejudice the evaluation of alternatives.

In a similar case, the owner retained the services of a control system specialist to improve the process control because that "must have been causing the poor process performance." The owner's engineering group identified one control improvement as requiring study, the operations group identified another, and the control system specialist identified yet a third. However, the control specialist also recognized that the control improvements would not, in themselves, solve the process problems. Fundamental changes in the operation of the plant would be required, along with some major capital improvements in the process piping and equipment.

Although neither case relates directly to digital control systems, both point out the need to acquire the proper talent for the planning process. In many cases, this will require several different specialties. In the first case cited, a general engineering consultant and a control system specialty consultant were used to evaluate both the mechanical and the control improvements after the first attempt proved not totally successful.

Planning Problem #2 — The Cloudy Crystal Ball

In this case, the owner decided to reduce manpower costs by remotely controlling some pumping stations. This proved to be quite cost-effective and, over the years, more and more remotely controlled facilities were added to the dispatching system. As the system was expanded, a computerized monitoring system was added to generate alarms and acquire data for long-term analysis of additional improvements. Everything seemed to be working fine. Fewer operators were needed at remote sites, and the efficiency of operation seemed to improve steadily. However, as the older operators who had grown up with the system began to retire, it became apparent that an excessively long training period was required for new personnel. In addition, it was found that operators were not effectively controlling all the facilities at their disposal. They tended to concentrate on the more important areas and to neglect those areas thought to be of secondary importance. Because of the continuing system expansion, the operators were becoming overloaded with information and control actions.

What started as a very effective way to improve operation and reduce costs quickly became an operational nightmare. Several operator mistakes caused equipment damage and loss of service. The computer was rarely used during the most critical operating periods because the operators did not have time to observe the data it was displaying and still perform the large number of manual control actions required. In addition, the panels were so large that an operator could no longer effectively coordinate the operation of the facilities.

When centralizing control, planning must address the ultimate objective of the control center. It must consider the operator's ability to absorb information and react to the data presented [7,8]. It must consider the implications of long-range expansion of the system and the ability of the operators and the control system to adequately execute the control functions. The digital control system must provide an integrated system.

Planning Problem #3 — Computer Reliability

An owner wanted a central operating facility for a large plant. Studies showed that a computer control system would be a cost-effective alternative for the large analog control panel originally envisioned. In addition, some of the control loops were quite complex and depended on a number of instruments and control devices.

A dual central computer control system was specified. One computer would back up the other in case of failure. System availability would be about 99% with this configuration. However, despite the high availability, the decision was made to provide full analog control backup at various panels throughout the plant. Many of the analog control loops were extremely complex and required the routing of a great number of signals around the plant site. It turned out that few of the analog controls were used. Portions of the analog system never worked due to their complexity. Modifications to the analog controls were extremely difficult when the plant was expanded. Because many of the control panels were scattered throughout the plant, it was difficult for operators to judge the effect of their control actions on the entire plant. The central computer facility was the only place where the entire plant operation could be observed.

In this case, the analog controls attempted to duplicate the computer controls. In many cases, this was not necessary. Many of the control loops could have been simplified to perform critical backup functions, while leaving the more complex control functions to the computer. A significant amount of money was spent for very little benefit.

Planning must address the control philosophy of the plant. The amount of automatic backup control required is related to the nature of the plant. Certain processes may require automatic backup. Others may operate well for the short periods when the computer is unavailable under manual control modes. Each control loop must be analyzed for the backup requirements. In addition, the location and amount of information displayed on backup panels is critical. If the panels are not easily accessible or do not display the proper amount of information, they will not be effective in providing the level of backup control desired.

Planning Problem #4 — Avoid Manual Usage

In this case, the owner decided that if backup control panels were provided, the operators would not use the computer control system. In this installation, control is either through the computer or from control stations located at each device. This is the opposite extreme from Case 3.

In some process areas, this approach to control was adequate. When the computer failed, the operators were able to control the process adequately from the various devices. Radio communication assisted in the operation. However, a few of the control loops were for flow splitting. It was found that it was almost impossible to operate the facility when the computer was not available. The owner was forced to install a monitoring panel near the center of the tank farm where the flow splitting was performed. In this way, one operator could observe the impact on control while another made control adjustments.

Several owners have expressed a similar concern that operators will not use the control system if local manual backup panels are provided. This seems to be a management rather than a control problem. Control systems must be designed to provide the proper level of control under failure conditions. They should not be designed to overcome management shortcomings in their enforcement of plant operation requirements.

DESIGN

Typical Approach

Design approaches will vary depending on the size and nature of the project [9,10]. Small projects generally have a relatively small design team. Specialty projects such as retrofitting a computer control system to an existing process generally have a design team composed of a few engineering specialties. Owner involvement in retrofit projects is usually heavy. The preceding types of projects do not have as many potential problem areas as do the large, interdisciplinary design-type projects.

Large design projects usually require services of a number of different engineering disciplines. These disciplines may be from different divisions or departments within a company or may be from different consulting firms. As more personnel are involved and as the project generally consists of various specialties, there is a greater chance that coordination problems will arise. For both large and small projects, the design is usually based on preliminary planning documents. These documents may have established the basis for design or have set the plant operating philosophy to be followed. Additional design coordination documents may have been developed as the first step in the detailed design.

For almost all design projects, the classical approach is to do the process design followed by the mechanical/structural design and, finally, the electrical, instrumentation and controls design. This approach works well in most classical designs. However, when applying previously untried process technology it can lead to significant problems. It can even lead to problems in relatively straightforward designs as discussed in design problem #1.

Most of the design problems described in the following paragraphs can be traced to a lack of early and continuous involvement of the electrical, instrumentation and controls personnel. Many mechanical and structural designs were compromised because of changes required by the controls personnel at a late phase in the design effort.

Design Problem #1 — Fuzzy Interfaces

One of the greatest areas for coordination problems is the interfacing of the instrumentation and controls to the mechanical and electrical equipment. In one case, the motor control centers specified by the electrical designers did not match the control panels specified by the control designers. In this instance, the controls and process designers worked closely during the design effort. The electrical design was done very late in the project, and the electrical designers were not involved in the process operational philosophy discussions. As a result, they misinterpreted the process design intent.

In a related case, the design philosophy required that device protective interlocks be furnished with the mechanical equipment. Process-type interlocks were to be performed by the control system. The mechanical designers used an off-the-shelf specification for some screening equipment. This specification contained both device-protective interlocks and process interlocks. As a result, the control panels and computer control strategies did not match the mechanical equipment operation. This problem was caused by the mechanical equipment designers not being involved during the process and controls design philosophy discussions.

Design Problem #2 — Supervendor

Many designers have been victimized by supervendor. Supervendor can solve all problems with his equipment. No matter what design constraints the engineer may point out to supervendor, the answer is always "no problem." In the area of computer control systems, the designer may work closely with vendors to ascertain the latest state-of-the-art in their product line, which is necessary because the art changes rapidly in the digital control field. However, if the designer relies overly on vendor inputs (especially those from a single supervendor), significant problems may result.

A designer worked very closely with a vendor during the design of a centralized computer control system. The vendor had many inputs to the

CAVEAT EMPTOR – LET THE BUYER BEWARE   125

wording of the specifications. Unfortunately, this vendor was underbid by another vendor. Because of the specification wording, the low bidder did not fully understand the scope of the work. As a result, the project was completed four years behind schedule and there were many claims during the construction phase.

As no two vendors offer the same hardware configuration, the hardware specifications are of secondary importance in the project. The functions to be provided should be as detailed as possible. Prospective vendors then can tailor their hardware to the functions desired. A common mistake among designers is to pick the best functions of several vendors and combine them into a single specification that no single vendor can meet. This can only lead to problems during the implementation because the vendors do not want to provide custom modifications to their standard features. The vendors realize that any time they must modify their standard packages there is a risk that something will go wrong and their costs will increase.

As a manager, care must be exercised in the selection of designers. Personnel must have the knowledge to realize the importance of filtering vendor inputs to avoid potential problems. In addition, as each vendor is different, managers must realize that it may be in their best interest to compromise with the vendor during construction if his equipment does not match the specifications exactly. It is better to obtain a standard system that the vendor will support at the sacrifice of a few functions that are not essential than to risk delays and claims by forcing the vendor to modify standard features.

Design Problem #3 — Undefined Requirements

Some computer control specifications appear to be little more than a set of one-paragraph control loop descriptions and a lot of hardware specification. Very little was stated about the intended system operational description. This describes how the computer hardware pieces were supposed to fit together to form an operational system and what should occur if something were to fail. In addition, there was little on the operator interface definition as to what displays were required and how they should be accessed. In this type of specification, the user ends up with little more than an analog replacement, which certainly would underutilize the capability of the computer system.

One of the major causes of this situation is lack of sufficient design time or dollars. Most architectural and engineering fees range from 6 to 10% of estimated construction cost. It is not uncommon in control systems for the design to range between 20 and 200% of construction, depending on the level of detail in the specifications and the complexity of the controls. Many managers do not realize the cost of the system engineering because it is buried in construction costs. Up-front design is costly but will reduce construction costs significantly. Two other causes of the lack of detail are inexperience on the part of the digital control system designer and vendor-dominated specifications.

The end user of the system will pay for the detailed systems engineering in one form or another — during design, construction or possibly in court. It seems more logical to pay for this work at the design stage of a project, when events are more under the manager's control.

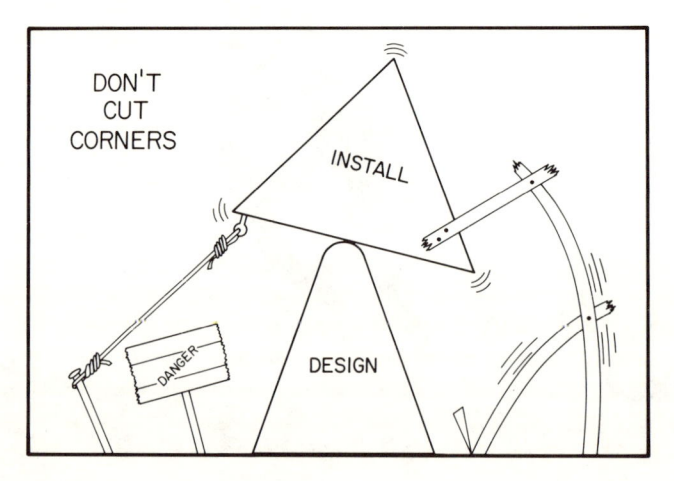

Design Problem #4 — Titan Missile Complexes

An owner wanted to control the sludge withdrawal from a reactor/clarifier. The process consisted of one backup and two online variable-speed pumps.

The designer selected a microprocessor-based control system with a VDT display located at the control station for the reactor/clarifier. The VDT was to be used to allow operators to control the process and to display alarm conditions. Data including alarms and process control status were transmitted to a central control computer. A separate panel located approximately 10 feet from the VDT contained a 100 window alarm annunciator and full backup analog control system. It is unclear whether this was a contest between the digital and analog designers for the most elaborate system or whether the owner did not trust either analog or digital control systems. In either event, it certainly seems like overkill in the control system design. The control system was far too complex for the simple process it had to control. A digital control system with a manual backup panel certainly would have been sufficient.

Design Problem #5 — The Bigger the Better

A sand filter system was designed to include the addition of air during the backwash sequence. The process designers and mechanical designers selected an air compressor based on the required air flowrates. Two compressors were specified: one as the online unit and one as the backup unit. The process required a fairly large air flowrate, and the compressors were sized with 150-horsepower motors. The structural designers then proceeded to design the foundations and room for the air compressors. Subsequently, when the electrical and controls group began their design, it was noted that the compressors could run cyclically with a four-minute on time and a one minute off time during certain process operational conditions. This is not recommended practice for motors of this size.

As it was too late to redesign the structural and architectural portion of the work, the controls and air piping were modified to allow the compressors to run continuously with excess air being vented. Had the electrical and controls group been involved during the earlier portion of the project, this problem might have been avoided.

Many designs have had to be changed late in the project when the electrical or controls group began studying the process dynamics or possible failure/process upset modes of operation. The equipment originally

128 PROCESS CONTROL COMPUTER SYSTEMS

envisioned will work well under normal operating conditions; however, its performance may degrade significantly under unusual conditions.

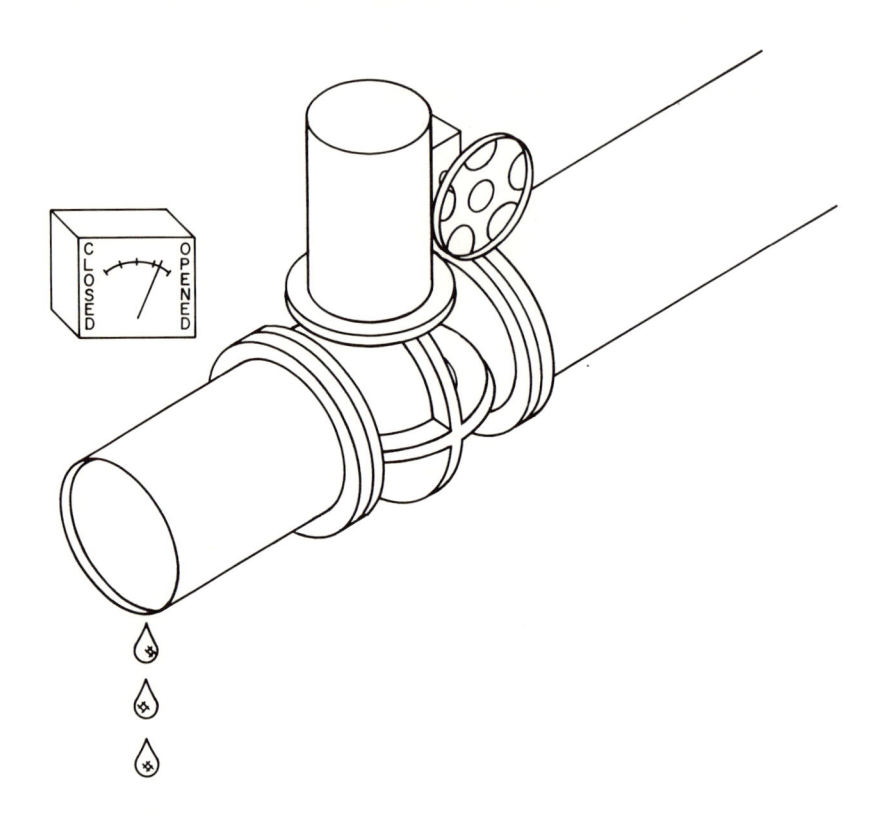

Design Problem #6 — Too Small for Design

A common problem is evidenced in applications that are felt to be so small and straightforward that the user goes directly from planning to implementation. This only works if the job can be conceived, planned and implemented by one person, who is also the user. That person's expectations have a chance of being realized.

Unfortunately, this is generally not the case, as, for example, with the food processing company that wanted to install a batch control system in one of its plants. The system was to control mixing of a number of different baking mixes, according to a set of programmed recipes. The system was conceived by the operating personnel, "designed" by a corporate engineering group and supplied on a turnkey basis by a vendor.

The "design" phase was little more than a series of verbal sessions between the engineer and the vendor, resulting in a purchase order. Unfortunately, the results proved this statement to be wrong. The results were missed budgets, missed schedules and a vendor who ultimately had to modify the computer physically to make the system work. The following were found:

1. The application was straightforward but the vendor did not know or understand it, although he was doing the programming.
2. The engineer understood all the complexities, particularly of the control software required.
3. The field interface was far from straightforward.
4. As there was no detailed specification, no one had a "blueprint" against which progress could be measured. Everyone had a different concept of what the result should be.
5. Due to lack of involvement, the plant people did not like what they received. It did not even come close to their expectations.

A detailed, written design and specification, even for small projects, is imperative. It tells the programmer exactly what is required. It forces the designer to consider all the interfaces and subtleties of the control. It provides a written "model" that can be continually compared to the product during development, and it serves as a concurrence vehicle for everyone, a way to be sure that everyone envisions the same result.

IMPLEMENTATION

Typical Approach

There are a number of methods utilized to acquire and implement a digital control system. The user may buy the basic hardware and software and perform the installation and application programming using in-house or consulting staff. The system may be purchased as a complete package, including vendor installation and application programming. The purchase may include only the computer control system or may include the computer and all field control panels and instrumentation.

Lastly, the computer controls may be acquired as part of a plant construction contract.

Vendor selection may be based on prequalification procedures, low bid, negotiated bid or as part of a design/build contract. Once the vendor has been selected, there is normally some monitoring of progress during the period when the vendor is designing, programming and testing his equipment at the factory. This monitoring may take the form of progress reports prepared by the vendor, shop drawing submission and review, periodic factory visits to monitor progress, or full-time factory observation by the owner or his representative.

Implementation Problem #1 – "Who's On First"

To circumvent some of the construction problems, or because of the proprietary nature of some industrial process, owners have acquired the digital control system through purchase orders. The hardware is purchased from the vendor, and the owner does the installation and applications engineering. The vendor may be used to assist both in writing the initial purchase order and in the initial installation of the hardware.

Many systems acquired in this manner have been implemented quite successfully; however, the owner must assume some risk. In most cases, the details of the application engineering have not been fully defined prior to purchasing the equipment. As the applications programming is performed, the user finds out that he does not have the proper "tools" to perform the work. These tools may include: the programs necessary to facilitate implementation of the application work, computer size limitations, poor vendor documentation on how to use the tools, or subtle system limitations that were not noted at the time the hardware was acquired.

Hardware must be carefully selected by knowledgeable personnel. Over-reliance on vendor salesmen can lead to significant problems at the implementation stage. Honest underestimates on either the part of the vendor or the application programming staff must be accepted. Corrective measures must be taken in a good faith effort to complete the work.

Implementation Problem #2 – Rubber Stamp Approval

To design and implement a control system requires a great deal of coordination among the various engineering disciplines. For example, there are at least three common methods to control a pneumatic valve used for flow control. Within these three categories, there are numerous variations due to location of control panels and the control philosophy selected.

Assuming that the mechanical, electrical and control specifications have been coordinated, care must be taken in the review of various shop drawings to ensure that each supplier has furnished the proper equipment. In many cases, several years may have passed between design and implementation. Different personnel may be assigned to the project. It is easy to forget why a particular valve actuator was specified with a certain type of control.

For control systems installed as a part of a construction contract, there are typically mechanical subcontractors, electrical subcontractors and the digital control system subcontractor. Numerous equipment suppliers may be submitting cut sheets on proposed equipment. Timely review of contractor submittals is critical so progress is not impeded. Field changes may be required due to unforeseen site conditions. Every submittal and every field change has a potential impact on the control system.

In one case, the electrical subcontractor furnished motor control centers that were incompatible with the computer control system. The electrical subcontractor made some incorrect assumptions in the preparation of his bid. The individual who reviewed the electrical shop drawings did not thoroughly consider the control system and approved the submittal. After the subcontractor ordered the motor control centers, the disparity was noted by the controls subcontractor. Numerous delays were encountered while the various parties attempted to solve the problem.

In another case, a pipe gallery was modified. This required relocating some piping, which then did not provide the proper upstream and downstream flow conditions for a critical plant flowmeter. As a result, the control had to be modified to place less reliance on the flow reading. This was a field change approved by the resident inspector, who did not thoroughly investigate the impact of the change on the controls.

In a third case, a supplier was to furnish a complete package process, including control panels. The computer system was to interface with the process for monitoring and limited control. Between the time of design and construction, the operation of the packaged process was changed by the supplier to enhance operation. He submitted the enhanced version of the process for approval and stated that it was different than specified but would be more efficient. This was approved by the mechanical equipment reviewers. Major changes were required to the specified control. The rubber stamp approval without reference to coordination with the control system subcontractor allowed the general contractor to escape liability.

The control system is the nerve system of the plant. Even supposedly insignificant changes can affect it. A small cut, if left unattended, can lead to much pain and anguish later on in the project.

Implementation Problem #3 — Dirt and Concrete Inspection

A small treatment plant was constructed and included a computer control system. The design consultant reviewed the shop drawings. The owner performed the field inspection duties. The plant was completed and started up with some "minor" punchlist items. The computer control system contractor arrived onsite to check out the controls shortly after startup of the plant. It turned out that 75% of the field instrumentation and controls were not working and had never been checked out. It was claimed that the checkout of the field equipment could not be performed until the plant was in operation and the water was flowing through the plant. Since there was some manual control capability on some of the pumps, all other instrumentation and control problems were considered minor by the field inspectors. After all, flow could go through the plant!

The owner assigned an electrical engineer and two maintenance technicians to the plant to resolve some of the problems. Over the next two years, few improvements were made because most of the problems were on a "punchlist" and, as such, were still the responsibility of the contractor who was no longer onsite. Finally, the plant manager dismissed the construction inspection group and had the problems corrected out of his budget. This was after several major process upsets had occurred due to lack of proper monitoring and control equipment operation.

Unfortunately, this is not an isolated case. The inspection groups at many plants seem to consider the instrumentation and control systems as secondary. As long as all of the dirt and concrete are placed properly and some limited manual control is possible, the plant is ready for startup and operation. Many inspection teams have somehow convinced themselves that controls cannot be checked out prior to operation. This is nonsense! Signals can be generated at instruments to check wiring continuity. Control panel functions can be tested up to the motor control centers and final elements.

This type of testing takes time and money, and the contractor will resist it unless it is specified. Many field signal generators and other test equipment are required. Electricians must be available to disconnect power and instrumentation wiring from the devices. Records of the detailed checkout must be kept. Electrical designers must be available to analyze problems and suggest changes.

The time and money can be spent before startup or after. After startup, one also has to worry about the potential impacts of testing on the process. In addition, there will be some parts of the control that cannot be tested until after startup. Problems here certainly can be compounded if one is uncertain whether the instrument is faulty or whether the wiring between the instrument and panels is faulty.

Implementation Problem #4 — Vendor Intimidation

When problems arise on a construction project, vendors can be very intimidating. In the area of digital control systems, we hear such comments as: "That's not my standard way of doing it," "The specifications are deficient," "That's not my fault," "You prove I'm wrong," etc.

Vendor intimidations are only a symptom of the problem. The vendor probably has underestimated the scope of the work and does not have sufficient monies remaining to furnish what the owner has specified or thinks he has specified. The owner does not want to spend additional monies to have the vendor provide something to which the owner thinks he is entitled.

It is obvious at this point that there is no meeting of the minds on the part of the owner and the vendor. The vendor may not have understood the owner's intent for the control system, while the owner may not realize the true impact of the vendor's equipment limitations and the implications of forcing the vendor to modify his standard.

Prequalification of prospective vendors prior to accepting bids can help each party come to a meeting of the minds prior to construction. However, there is usually no provision in prequalifications to reimburse vendors for the time spent adequately addressing the functions of the specifications. As a result, most prequalification packages prepared by the vendors are limited in scope and general in nature.

A second approach to obtain an early meeting of the minds is to prequalify and then to negotiate a bid price. This allows both parties to understand the other. A third approach is to jointly develop a work plan and work scope shortly after awarding the contract. The specifications are then replaced or supplemented with the negotiated work scope.

Implementation Problem #5 — The Never-Ending Punchlist

The never-ending punchlist is an owner tactic to avoid final acceptance of the computer control system. The use of this tactic can have several causes. The owner may not be confident in the operation of the computer or may not have a trained operations staff available. Perhaps all the functions cannot be tested due to field problems. Management may not know what to do with the computer or may fear making a bad decision. For whatever reason, it seems that the owner continues to find few reasons why the system cannot be accepted. Vendors are quick to counterattack by demanding final punchlists, abandoning the work or threatening claims and litigation.

One owner was able to delay acceptance for three years. Another ended up in court within a couple of months. In both cases, neither owner was totally pleased with the system as accepted and neither has acquired the proper operation and maintenance (O&M) staff. Both owners eventually made modifications to the systems to better suit their needs.

The manager must accept that he may not like everything about the control system at acceptance. The manager should ensure that the specifications require the vendor to furnish the tools required to allow changes to be made. And, most importantly, the manager should acquire the staff to operate, maintain and optimize the system.

STAFFING

Typical Approach

The first step is to determine what staff is needed. This is not easy for a manager who may not know exactly what kind of staff a computer control system requires.

Of course operators will be required. The question is what kind of skill level is required and how many will be required? How much work will the control system do and how much will the operators have to do? Will the control system be trusted to operate the plant on the off shifts and thereby reduce the number of operators required?

Will the staff maintain the computer hardware or is a service contract desired? Who will make changes to the control system programming? Who will manage the control system?

Once these decisions are made, one may have to develop job descriptions, salary levels and tests. Union negotiations may be required. Training programs may have to be established if in-house personnel are to be utilized. More about staffing will be discussed in a later chapter. Presented in the following sections are some case histories of some potential problem areas.

Staffing Problem #1 — Rose-Colored Glasses

In a large plant, the owner decided he would do all work associated with the computer system. This included training control room operators utilizing his existing training staff, maintaining the computer control hardware, programming and process control application engineering. The construction period was about two years, and the owner began to consider staff acquisition and training about six months after the start of construction.

The operator job title creation and union negotiations with the existing operator union required almost twelve months before everyone agreed to duties, salary and the selection process for control room operators from among the existing operating staff. While operators were being identified, a training program was established. It became apparent that the existing training staff knew nothing about the computer control system being furnished. This was solved by hiring a consultant to train the first group of operators along with the training staff. Time passed.

Process engineer talent was being developed during the construction. However, six months into the project, the engineer-in-training left to accept a better job offer. Six months later, another engineer was hired to take his place. Time passed.

Additional maintenance personnel were required. These were hired under instrumentation technician titles and were sent to computer maintenance classes offered by the vendor. Computer systems can be quite complex and unless one works with them continually, one tends to forget some of the more subtle problem diagnosis and repair techniques. Time passed.

The computer was delivered and started up. The vendor was responsible for some operator training. He also performed all hardware maintenance. The operators did not learn much from the vendor training. The maintenance personnel were assigned to other areas of the plant because the vendor was performing computer maintenance. Time passed.

The owner now has a maintenance contract with the vendor. Operators were given on-the-job training by the design professional who assisted in the startup and checkout of the computer control system. A programmer and an electrical engineer were assigned to assist in the diagnosis and correction of field and programming problems. The design professional is doing the modifications to the control system application software, and no one is officially in charge of the control room or managing the computer resources.

This owner had previous experience with computer control systems and was relatively well prepared to staff the system. Despite his efforts, several unanticipated problems, such as staff turnover and union problems, slowed his implementation. Perhaps his program outlook was too optimistic. Long lead times are required to acquire and train a staff. Rather than planning major reorganization all at once, a more incremental approach could have been taken.

Staffing Problem #2 — Money

It can be very difficult to convince upper management that staff should be acquired several years before a project is scheduled for completion. It also raises havoc with operating budgets. In addition, political or public opinion may have a strong influence on early staff acquisition.

In one municipal plant, the plant manager was not allowed to acquire operating personnel until a facility was ready for startup. This was an expansion to an existing plant, and the theory was that existing operators could cover both the old and new plant areas while new operators were acquired and trained. The plant was able to proceed until a computer control system was installed. In previous expansions, the operators could relate their past experience to the new areas. However, they had no experience with computer control systems. Needless to say, the computer systems were very poorly utilized for the first 18 months while the operators learned how to use it effectively.

A second monetary consideration is the problem with unions. Many computer control systems are no more complex than the analog controls they replaced. In fact, in many cases, they are actually easier to operate. However, the word "computer" control seems to arouse the best in union negotiators to demand higher salaries for their members. As management may not be familiar with the details of the computer operation or have even considered how it will fit into the plan of operation, it may be very difficult to defend a "no pay increase" position.

If the manager believes he may be faced with monetary limitations or union problems such as described above, there is a solution. It is usually possible to have one or two people added to the budget. If possible, during the design phase of the project a technical person should be assigned to become familiar with the design philosophy of the control system. As construction progresses, the individual should learn as much as possible about the control system operation and use. The key person can assist management in setting up a plan of operation for the computer controls and in union negotiations. A few months prior to startup, he can assist in training operators and can be utilized as a chief operator until sufficient operating staff is acquired and trained. Subsequently, he can function as the manager of the control system in coordinating operation, maintenance, training and optimization of the control system.

Staffing Problem #3 — Overqualification

An owner did not believe that his existing operators could be retrained to operate the computer control system. Instead, he transferred personnel from the lab staff. One requirement was that the operator be a college graduate. Most of the personnel were chemists by training. The theory was that any college graduate should be able to understand the computer control system. This assumption was correct. Most operators had no trouble operating the computer controls. Unfortunately, they had no comprehension of the process. Field operators would call with process questions and they could not answer them. As a result, the field operators would take over from the control room staff because they had lost credibility with the field operators.

There was one exception to the above. One control room operator was not a college graduate but had a lot of field experience at the plant as an inspector during the checkout of the computer. He understood the functions of the computer as well as the operation of the field equipment. He is now the head control room operator. He has credibility with the field operators and has taken night classes in the theory of operation of the process.

Staffing Problem #4 — Transfers

One plant planned to staff the control room with existing plant operators. When the job was posted, many senior operators refused to apply for the position despite the salary increase offered. They were accustomed to operating from the field and being within sight and sound of the equipment they were operating. They did not like the concept of remote operation. Other senior operators who did accept the position later requested demotion back to the field.

In the end, many of the most experienced operators were in the field while the control room operators were less senior personnel. When control room personnel would request field changes, some senior operators were upset by the young upstarts in the control room giving them direction. Some even were able to convince the control room staff to countermand their directives. On some shifts, there was a real problem as to who was running the plant — the control room operators or the field operators.

Staffing and training are much more under the direct control of the plant manager than are the design and construction activities. This is the area that can make or break a control system. A good staff can make a poor control system function to its peak capability. A poor staff can make a good control system function poorly. Proper management of the staff is critical if the control system is to be a success and is to be optimized to provide the best possible control.

MANAGEMENT AND OPTIMIZATION

Managers have risen to their positions either through the ranks of the company or through experience gained at a similar organization. In general, they know the process and the plant equipment, the plant staff and the organization and operation of the company.

As managers move up in the organization, their knowledge base expands. Managers know how to accomplish the organization's objectives through managing people, although they may not know the exact details of how each person performs his duties to accomplish the overall objectives.

One day, the engineering staff or upper management informs the manager that the company is to become the proud owner of a computer control system. The manager may have little or no knowledge about a computer. Even worse, the manager may have no idea what to expect and how the control system is to be managed.

Management Problem #1 — The First Computer Syndrome

After being told that the computer will not only control the process but that it can generate operational reports and provide maintenance schedules, among many other things, there seems to be an overwhelming tendency to give up or take classes to learn everything there is to know about computers. Perhaps other plant managers have related the problems associated with computers, or perhaps the television commercials about home computers have finally shown convincingly that this is the best way to learn about computers.

If a new piece of equipment or a new process is installed at the plant, does one immediately go out and learn everything there is to know about the equipment? Does one give up? Most likely, this task is assigned to subordinates. The manager's major concerns will be how many people will be required to operate it, what will it do to the operating budget and what it is supposed to do. Why the difference with a computer system? Perhaps there is no subordinate who is knowledgeable about computers. Perhaps this is a management level function.

Do's

1. Do acquire at least one staff member who can assist in managing the computer as soon as word is received that a computer control system will be furnished.
2. Do attend a seminar or two on computers. If you can find one on management, attend it.
3. Do require your design and construction groups to work with your staff throughout the project.

Don'ts

1. Don't assume you can learn everything about computer control systems.
2. Don't assume you will have time to learn anything about computers.
3. Don't assume you will be able to manage the computer control system without some assistance.
4. Don't pass the responsibility to some subordinate who has never seen a computer control system and then expect immediate answers.

Management Problem #2 — Employee Fears

There is a natural tendency to fear the unknown. Experienced operators suddenly start losing sleep over "the computer." Many managers, especially those who may have experience with computers, tend to overlook potential employee fears of the system.

Employees must be kept informed of the status of the computer work. Seminars should be given well in advance of the arrival of any new control system to alleviate fears. How the system will work and what the operators will be expected to do should be discussed freely.

At one plant, operators were given regular seminars starting about six months before the anticipated delivery date of the computer control system. These seminars included discussions of typical operator duties, the theory and operation of computers, what to expect during startup and checkout, what kind of problems might result during normal control by the computer and what differences might exist between the way the plant was currently operated and how the operation would change after acceptance of the computer. Operators also were primed on what type of questions to ask the vendor during training and checkout.

When the computer arrived, the operators understood what it could and could not do and how important they were to ensuring the success of the system. After a week or two of becoming familiar with the mechanics of operating the process through the computer, they performed admirably. Very few problems were encountered with operator complaints about the system, except when it did not operate properly.

Management Problem #3 — Who is in Charge?

An existing plant added a central monitoring and control computer. Prior to the addition of the computer, the plant had been operated on a unit process basis. There was great difficulty in coordinating the operation. It was expected that the computer would solve some of these problems by centralizing the plant operating data display and providing better management data. The existing field operators remained in their previous positions. Less-experienced operators were assigned to the control room.

After several months of operation, little improvement was observed in the coordination of the unit process operation. Many devices were in local control, not computer control. Management reports were based predominantly on field reports, rather than central control room reports.

The field operators had sensed a potential loss of their power with the installation of the central control facilities. There were some cases of sabotage of equipment, and there was very little cooperation between the field and the central operators. Little or no information was relayed from the field to central control. In some cases, the information relayed was misleading or erroneous.

The managers of the plant did not insist on management reporting through the central control room, which further undermined control room authority and gave credence to field operator contentions that they were still in charge of the operation. There were no clear written guidelines as to the authority and responsibility of the central operators and the field operators.

Over a period of about one year, many of these difficulties were resolved. Field operators were given responsible assignments in the area of custodial maintenance, emergency operation, equipment status reporting functions and coordination of maintenance activities with the central control operators. Central control operators were responsible for the process operating parameters, such as flow control, and were responsible for preparing management reports.

When new control philosophies are implemented in existing plants, great care must be taken to clarify new lines of authority and responsibility among the operating and management personnel. No one likes to feel that he has lost power over his areas of concern. Personnel must be convinced that their efforts are being redirected for the good of the overall operation.

Management Problem #4 – The Paper Process

How many operators in your plant have read each operation and maintenance manual? It has been my experience that most O&M manuals are too voluminous for most operators. They are interested in operating the process, not reading about its operation.

Computer system O&M manuals are typically too complex and too voluminous for the average operator to utilize effectively. Most manuals are oriented toward individual pieces of equipment, while the control system is oriented toward the process interactions. Most manuals are weak in the areas of defining duties among the shifts and personnel on each shift; in defining reporting functions, coordination procedures and follow up; and in defining contingency operation in terms of the personnel involved.

The lack of clear, concise manuals and operating procedures can result in operators taking the wrong actions, taking no action or taking an excessively long time to determine the correct action to take. It also can result in poor coordination among personnel. No one is sure exactly what to do during both normal and emergency conditions.

Control system manuals are difficult to write. A computer control system may perform a multitude of functions. It is difficult for the operator to remember all the functions. Contingency control responses occur infrequently. When they do occur, the operator may not know what to do. In one case, the computer was programmed to stop all control of a bar screen if the rake mechanism failed to stop within a certain time period after the controls had issued a stop command. This time period was the time it took for the rake to reach the "home" position. Control worked fine for several months. One day the control program stopped. The operators could not understand why this occurred. On investigation by the control program designer, it was realized that the time period programmed into the computer for the rake to stop was about one minute too short. If a stop command was issued when the rake was at its lowest point, the stop period would be exceeded.

This condition was described in the O&M manual. However, it was buried in the 15 or 20 pages used to describe exactly how the computer

142 PROCESS CONTROL COMPUTER SYSTEMS

was controlling the bar screen operation. Subsequently, each computer control program function was condensed into a one-page summary that described what should happen under both normal and abnormal conditions. It described how the operator could detect problems and what to do about them.

Additional one-page summaries were developed for other tasks the operators were expected to perform. This included when to have the computer print operating reports, what shifts were responsible for checking the validity of monitored field sensing devices, etc. Each operator has his own copy of the one-page summaries and is allowed to organize the manual any way that is convenient. The operator is encouraged to suggest modifications and to make notes on the sheets. The operators are creating their own manuals.

SUMMARY

Caveat Emptor

Many problems can arise from the initial planning of a computer control system to the operation and maintenance of the completed system. Many can be minimized with proper management of the project. This starts with the acquisition of the correct talent for the planning and designing of the control systems. It ends with good management of the properly trained operating staff. All these functions are under the control of the managers.

Successful Systems

Successful systems are built on a good foundation constructed during the planning and design phases of a project. On this foundation must be placed the proper implementation and staffing. The mortar that holds it together is the knowledgeable management of project personnel.

The chapters that follow describe the success steps that can be followed in the planning, design, implementation, staffing and management and optimization of the computer control system. In reading these chapters, one may consider some of the problems highlighted in this chapter and how they could have been avoided.

REFERENCES

1. Hegg, Bob A., et al. "Evaluation of Operation and Maintenance Factors Limiting Municipal Wastewater Treatment Plant Performance," U.S. Environmental Protection Agency, Report No. 600/2-79-034 U.S. Government Printing Office, Washington, DC (1978).
2. Water Pollution Control Federation. "Operation and Maintenance of Water Pollution Control Facilities: A WPCF White Paper," *J. Water Poll. Control* (May 1979).
3. Hill, W. R. "Why Treatment Plants Don't Meet NPDES Limitations," paper presented at the Indian WPCF Annual Meeting, November 9–11, 1981.
4. Rubinstein, E. "An Analysis of Three Mile Island," *Spectrum* 16(11): 33–57 (1979).
5. Ledgenwood, B. K. "Trends in Control," *Control Eng.* 120 (December 1980).
6. Steinmann, J. E. "The Vendor-User Relationship," paper presented to Lake Superior Section, Instrument Society of America, 4th Annual Symposium, June 19 and 20, 1969.
7. Sugarman, R. "Nuclear Power and the Public Risk," *Spectrum* 16(11): 61–69 (1979).
8. West, B., and J. A. Clark. "Operator Interaction with a Computer Controlled Distillation Column," in *The Human Operator in Process Control* (London: Taylor & Francis Publishers, 1974).
9. Skrentner, R., et al. "Design Approach for Automation of Activated Sludge Wastewater Treatment Plants," paper presented at the Instrument Society of America National Conference and Exhibit, Chicago, IL, October, 1979.
10. Manning, A. M., and D. D. Dobs. "Design Handbook for Automation of Activated Sludge Wastewater Treatment Plants," U.S. Environmental Protection Agency, Report No. 600/8-80-028 U.S. Government Printing Office, Washington, DC (1980).
11. Skrentner, R. "Management of Startup, Testing, Acceptance, and Operational Transition," panel discussion at the Instrument Society of America Annual meeting, Philadelphia, PA, October, 1978.

CHAPTER 7

SUCCESS STEP 1 – PLANNING

Rod Graupmann, PE

INTRODUCTION

The most basic management function is planning – the selection from among alternative future courses of action. Planning involves selecting objectives and goals and determining ways to reach them. Plans thus provide a rational approach to preselected objectives. Planning is deciding in advance what to do, how to do it, when to do it and who is to do it.

Planning for process control computer systems is as essential as planning for any other aspect of a business enterprise. Identification of system goals early in the project planning phase is especially important. Failure to clearly identify system objectives prior to procurement is, in fact, a frequent cause of a computer system's failure to meet user expectations. Only by careful planning can an organization reasonably expect that a process computer control system will satisfy all intended (and clearly identified) organizational needs.

The purpose of this chapter is to outline a logical approach to the planning process. Procedures for the following planning activities are suggested:

- Establishing needs and objectives
- Determining evaluation criteria
- Identifying alternatives
- Evaluating alternatives
- Developing an implementation plan

By following these steps in a logical and orderly sequence, the system user will be assured that the proposed computer system will meet his needs in the most cost-effective and beneficial manner possible.

WHY PLAN?

Without planning, business decisions would become random choices, as though a pilot set out without knowing his destination. Five specific reasons for planning are to focus attention on objectives, to offset uncertainty, to achieve an economical operation, to facilitate control and to gain management acceptance.

Focus Attention on Objectives

Because planning is directed toward achieving objectives, the very act of planning focuses on these objectives. Managers, being immersed in immediate problems, are forced through planning to carefully evaluate the future and identify needs and objectives.

Offset Uncertainty

All organizations are faced with an uncertain future. Processes change, new plants may be built or modified, personnel change and the outside world is constantly changing. These evolving conditions decrease certainty and make the planning process an important, albeit difficult, function. Assumptions must be made and plans formulated according to these assumptions. The plan must be flexible so that adjustments can be made when unforeseen events develop.

Achieve Economical Operation

Typically, business decisions are made on the basis of economics. Reduction of operating costs is often an important factor in a decision to procure a process control computer system. The planning process must include an analysis of capital cost, operating cost and resulting financial benefits to arrive at selection criteria such as a calculated return on investment. Cost/benefit analysis of alternatives during the planning process will increase the probability of reaching financial objectives, such as reduction of operating costs.

Facilitate Control

Predefined goals are necessary to later measure the performance of a system. The "fit" that the system's actual performance has with the plan's objectives will determine the degree of success that the system has accomplished. Objectives should be specific and measurable, rather than general, so that quantitative results can later be determined.

Gain Management Acceptance

Projects must be "sold" to management. A plan will provide the information necessary to sell the project if it is prepared properly. Involving upper management in the project planning process is an effective way to gain acceptance.

GENERAL PROCEDURES

The general planning procedure is illustrated in Figure 1. It consists of the following steps.

Initiate a Plan

Procedures must be developed to organize the planning process. The scope of the plan should be stated in writing. For example, will the plan address field instrumentation or will it be limited to digital system components? A schedule for completion of the plan, including intermediate milestones, should be developed. (Most plans without definite completion dates are never completed due to "lack of time" to work on them.) Manpower resources needed to develop the plan must be committed and a budget established prior to starting work. A task force may be formed to provide the necessary inputs to the planning process. A decision must be made whether to do the planning entirely in-house or to involve an outside consultant to assist in its development.

Establish Objectives

Objectives must be identified clearly in terms of both tangible and intangible goals. Tangible objectives are preferred. For example, it is

148 PROCESS CONTROL COMPUTER SYSTEMS

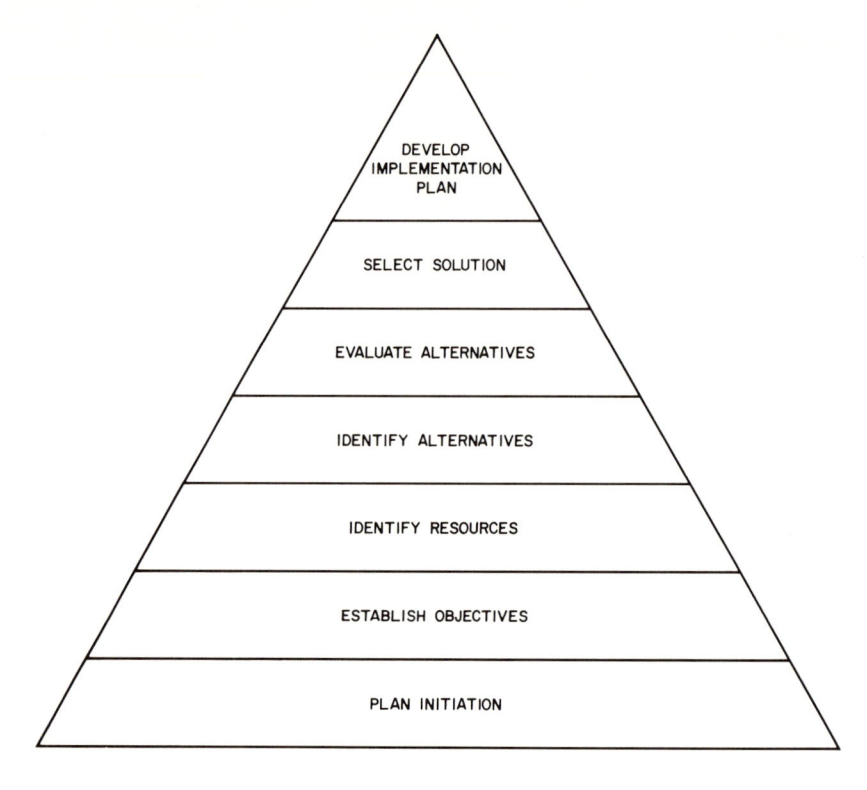

Figure 1. General planning procedure.

[Pyramid diagram with layers from top to bottom: DEVELOP IMPLEMENTATION PLAN; SELECT SOLUTION; EVALUATE ALTERNATIVES; IDENTIFY ALTERNATIVES; IDENTIFY RESOURCES; ESTABLISH OBJECTIVES; PLAN INITIATION]

better to state an objective as "increase plant throughput by 10%" than to simply state "increase plant throughput." Some benefits, however, such as increased control system flexibility, cannot be quantified and must be expressed in intangible terms. Objectives must be addressed in both the long and the short term. Long-term goals often reflect organizational policy, e.g., "increase plant automation to reduce product labor content." Short-term goals tend to be more specific and quantifiable, e.g., "reduce product labor costs by 10% during the next 18 months."

Identify Resources

A plan must consider the boundaries or limits of available resources. In any organization, both labor and capital are scarce resources. Upper bounds for capital investment must be identified to determine the "reasonability" of alternative system choices. Similarly, availability of personnel and the types of personnel available for system operations and maintenance functions must be identified. For example, if an organization must use existing operators for a new process control system, the system selected must be acceptable to those operators and compatible with their skill levels.

Identify Alternatives

Alternatives should be identified without prior prejudice. One procedure that may be followed to eliminate personal bias is to have a "brainstorming" session, involving all interested parties. During brainstorming, all ideas should be heard and listed without any attempt to discuss their practicality or desirability. The suggestions can be filtered and analyzed by the planning team following the session. Often, innovative approaches can evolve from such sessions. All "reasonable" alternatives should be documented clearly.

Evaluate Alternatives

The procedure that will be used to evaluate the various alternatives must be documented so that the analysis will be logical and well understood by all members of the planning team. Evaluation procedures can include mathematical as well as heuristic techniques. The alternatives are screened during this step, eliminating those that cannot meet the objectives set forth. Usually, some alternatives can be eliminated early in the analysis because their degree of "fit" with the stated objectives is negligible. The remaining alternatives will meet the system objectives in varying degrees. A rating system must be formulated to evaluate the feasible solutions.

Select the Solution

During this step, the feasible alternatives are compared using the rankings developed in the previous step. Cost/benefit analysis may be used to compare choices. The recommended solution is then documented along with the supporting rationale. The summary of the recommended system should include a brief description of all necessary hardware and software components, their associated capital cost and projected operational impacts and costs.

Develop Implementation Plan

Once the analysis is completed and an alternative selected, a plan should be developed to implement the system. The plan should include a schedule, cash flow requirements for budgeting purposes, manpower requirements and a summary of all tasks required to design and implement the system. Any auxiliary services required, such as architectural design for a new computer room or the use of an outside consultant for system design, should be addressed.

The final step in the planning process will be to assemble a complete report, including the results of the steps given above. A summary should be prepared, condensing the report to the important findings and recommendations for management review. Supporting documentation, such as calculations and the like, should be placed in a report appendix. It may be desirable to also prepare a presentation to "sell" the plan to top management.

FORMATION OF THE PROJECT TASK FORCE

Once a decision is made to develop a plan to procure a process computer system, a task force must be formed. The task force should be headed by a key member of management, who does not necessarily have to be a technician but should be knowledgeable in general system concepts and organizational goals.

Other members of the task force should include personnel knowledgeable in current computer system technology. An important point is that a process control system is a *real-time* computer system, which is significantly different from a data processing application. If in-house technical capability is not available, an outside consultant should be employed to provide the required expertise. One should not rely on data processing staff to provide all the technical input to a real-time control system study.

Plant operation and maintenance (O&M) also should be represented on the task force. The process control system will be the responsibility of these two groups once it is installed; it *must* meet their needs.

Once the members of the task force are selected, the project manager (task force leader) should propose a schedule and budget for completion of the plan. Individual task force member responsibilities and assignments should be documented. Regular task force meetings should be scheduled to coordinate the work and review each step in the planning process. A concise definition of the scope of the plan should be prepared and approved by management.

The task force project manager will be the key interface to upper management. It will be his/her responsibility to involve management in the planning process and to keep them updated on progress. Regularly scheduled briefings are a good method to obtain management concurrence each step of the way. These briefings should be concise and to the point. Rambling, disorganized presentations inevitably will be detrimental to project success.

Adherence to project schedule and budget is important. A plan that is completed on time and within budget will indicate that the plan can be implemented in the same fashion. The converse also will be true. Poor management of the planning process can lead to disapproval of the followup project, even though the ultimate recommendations may be sound.

NEEDS AND OBJECTIVES

The first task of the project team is to identify and document all system objectives, and the best way to do this is by using interviews. All key personnel who will use or maintain the system must be interviewed to identify their needs. The interviewer should be skilled in listening techniques, i.e., he must be able to ask good questions to draw out the user's true needs. The questions should be open-ended, such as the following:

- "What are your needs for operational reports?"

rather than

- "Do you need a shift report?"

The questions should be as open-ended as possible to avoid encouraging answers based on the bias of the interviewer.

At the end of each interview, the interviewer should summarize the stated needs. He should then ask the other person to prioritize, or weight, these needs as to their relative importance. For a plant process control system plan, the following personnel should be interviewed:

- Plant manager
- Assistant plant manager

152   PROCESS CONTROL COMPUTER SYSTEMS

- Operations (production) manager
- Maintenance manager
- Shift supervisors
- Maintenance foremen
- Selected operators

Other inputs to the needs identification process will be a thorough review of existing documentation, such as operational control procedures, O&M manuals, as-built drawings of existing control equipment, and the like. Field surveys of existing instrumentation also will be required for retrofit projects.

Once the interviews and reviews are complete, the task force will compile the needs list. The task force should reach a consensus on the system needs and then present its findings to operations and management personnel. Once general concurrence is reached, the task force can move to the next step in the planning process—establishing evaluation criteria.

ESTABLISHMENT OF EVALUATION CRITERIA

The output from the needs analysis will be a list of items ranked according to their relative importance. This ranking of needs will be used to evaluate alternatives.

One way to look at system needs is in terms of "musts" and "wants." Musts are those needs of absolute importance; any alternative to be seriously considered must satisfy these needs. Needs identified as "wants" are less than absolute and must be ranked, or weighted, as to their relative importance. The task force must establish the weighting system, subject to management concurrence.

The best way to illustrate this process is by example. A large municipality authorized a study to determine the feasibility and benefits of implementing a computer-based monitoring and control system for its water distribution system. The distribution system consisted of pump stations, storage tanks and interconnecting pipes. The needs analysis resulted in a list of needs identified as follows:

Musts

- Centralized data acquisition
- Centralized control of storage levels and pressures
- Centralized control of pressure regulating valves
- Centralized control of pump stations
- Centralized data storage and report generation

Wants   Weight

Wants	Weight
Additional sensors for pump malfunction alarms	5
Additional flowmetering	4
Additional pressure monitoring	8
Graphic display capability	6
Trend display capability	7
Data link to existing treatment plant computer	3
Equipment run-time monitoring	9
Power demand control	9

Notice that the "wants" were weighted on a scale of 1 to 10 as to their relative importance to the user. The musts were not weighted because they were all absolute needs, or "10s."

Assignments of weights to the list of "wants" may not be an easy task. During the interview, each person was asked to prioritize the list of "needs." Undoubtedly, these weights (and the needs themselves) will vary considerably among people. It is the task force's responsibility to reach a consensus on both the needs and their weights. It may be necessary to vote on each need if consensus cannot be reached through dialogue. The task force leader must provide the leadership necessary to reach final agreement.

For any system study, "limiting factors" will exist. These factors will establish ranges within which alternatives must fall to be considered. Capital cost is often a primary limiting factor. If the preliminary budget for a system is established as "$1 million maximum," obviously alternatives that cost $2 million should not be considered. Other limiting factors may include maximum implementation time, system reliability requirements, corporate standards for process control equipment and available O&M manpower resources. The evaluation criteria should include a list of these for use in the selection process. For example, the following list was used on the project discussed above.

- Capital cost less than $3 million?
- Implementation time less than 2 years?
- System reliability greater than 99%?
- Maintainable by existing staff?

The limiting factors can be added to the "musts" lists for the evaluation of alternatives.

As in any business decision, cost will be a determining factor in evaluating alternative process control systems. The evaluation criteria should include a definition of the basis for comparison of alternatives in terms of both capital and operating costs.

154   PROCESS CONTROL COMPUTER SYSTEMS

Common methods of economic evaluation include total annual cost, present worth, return on investment and payback periods. All require an estimation of initial capital investment cost, annual operating costs and annual operating savings. The following are examples of typical costs and savings included in an economic analysis:

Capital Cost

- Engineering
- Hardware
- Software
- Installation
- Environmental (floor space, air conditioning, etc.)

Operating Cost

- Hardware maintenance
- Software maintenance

Operating Savings

- Product throughput
- Manpower
- Chemicals
- Energy

The evaluation criteria must include a definition of costs and savings to be included in the analysis and a description of the method to be used to compare alternative courses of action, e.g., return on investment, payback period, etc.

In summary, the evaluation criteria should include the following:

- Definition of the economic criteria to be used for evaluation of alternatives
- Definition of the noneconomic criteria to be used for evaluation
- Weighted list of all control system "needs" and "wants" identified by the task force, with the "wants" weighted according to their relative importance

The evaluation criteria and method of evaluation should be presented to management for their concurrence prior to proceeding to the next step — identification of alternatives.

IDENTIFICATION OF ALTERNATIVES

Earlier in this chapter, the importance of brainstorming to generate alternative courses of action was discussed. The task force may wish to invite other technical members of the firm into the initial brainstorming session. Often, people with no prior involvement in the project will be more innovative in their suggestions because they are completely free of any prior bias or prejudice.

The task force leader should chair the discussion and write down all suggestions on a blackboard as they are generated. No ideas should be discouraged during this initial session. The leader should serve as a catalyst but should not attempt to influence the discussion towards his preferred course of action.

The initial filtering to be applied to the list is feasibility in terms of technical, economic and policy considerations. Many of the brainstorming suggestions will be discarded on the basis that they are not technically feasible for the problem at hand. Some may be too "state-of-the-art" to be considered for an industrial control application that requires high reliability. The company may have a policy, for example, that no systems may be considered if they have not been implemented previously for a similar application by other users.

The first screening will be based primarily on the "musts" list and limiting factors discussed in the previous section. Some alternatives can be discarded easily on this basis. Others will not be obvious and must be analyzed further. For example, if centralized report generation is a "must," distributed digital systems with no central computer or other report generation capability could be discarded in the analysis. If, however, report generation is identified as a "want," subject to economic analysis, a system without a central computer could be considered.

Once the initial list of alternatives is reduced by the first pass screening, the technical members of the task force should be assigned responsibility for a complete analysis of the remaining choices. The analysis should include the following.

- Research of each alternative, including literature review and contacting vendors to determine technical feasibility
- Discussion with other users with similar applications
- Estimation of system reliability for each alternative

- Estimation of the capital cost of each alternative, with capital cost estimates being verified with several different vendors
- Estimation of recurring maintenance cost of each alternative, including personnel requirements, instrumentation and digital system maintenance

Completion of the technical analysis provides a number of feasible alternatives and their associated costs. To compare these alternatives, resulting benefits also must be estimated. For process control systems, the following areas typically are addressed:

- Product throughput and quality
- Operational labor
- Fuel, power, chemicals and similar production costs

Product Throughput and Quality

The analysis should consider the present production throughput and the impact of alternative control systems. For example, if product rejections based on inferior quality can be reduced, quantifiable savings will result. Impacts on production line downtime also should be considered. Relative degrees of reliability can be translated into estimates of reduced downtime and resulting savings.

Operational Labor

Operational labor requirements should be estimated for each alternative. Often, automation is justified based on reduced operational labor; however, requirements may vary, depending on alternative control systems. For example, a decentralized control system with several operator stations may require more operators than a totally centralized system. Increased control system maintenance labor costs, including sensor calibration and cleaning, must be subtracted to calculate net labor savings.

Fuel, Power, Chemicals, etc.

The use of computer control systems to reduce power and/or fuel consumption is becoming increasingly popular as their costs rise at a rapid rate. A computer-based control system offers the opportunity to utilize advanced optimization strategies to minimize energy and chemical consumption in production processes. The savings that will result from implementing such a system are difficult to calculate exactly; however, they can be estimated based on theoretical calculations or savings achieved in similar applications by other users. The capital costs calculated previously must, of course, include the sensors and software required to implement the optimization programs.

Savings related to process optimization may vary by alternative. It may indeed be desirable to justify the implementation of advanced control strategies by cost/benefit analysis, i.e., calculate the cost of hardware, software and sensors required and then calculate resulting process cost savings.

Intangible benefits for each alternative also should be listed. These could be features such as increased flexibility, better man-machine interface, easier expansion, etc. These features are difficult to quantify in economic terms; however, a ranking system related to system needs can be used to evaluate alternatives, as explained in the following section.

EVALUATION OF ALTERNATIVES

The previous sections described development of evaluation criteria and alternatives. We now will discuss the procedures used to evaluate the alternatives against the criteria to develop the final recommendations.

Economic Evaluation

A cost/benefit analysis requires a comparison of capital cost vs annual savings with consideration of the time value of money. Because of interest, a dollar is worth more now than at any future time. Interest may be defined as the cost of borrowing money or as the return available from alternate investments. Generally, an organization will have an established interest rate, which should be used as the basis for economic evaluation.

A detailed explanation of the alternative methods of economic evaluation are beyond the scope of this text. Some methods will be summarized; however, for a complete explanation the reader is referred to a text such as the one by Grant and Ireson [1]. The first consideration in the economic evaluation is the summary of capital cost and annual savings developed in the earlier sections. Consider, for example, the following costs and savings for the water distribution control system discussed earlier in this chapter:

Capital Costs

Alternative 1	$2,500,000
Alternative 2	$2,850,000
Alternative 3	$2,225,000

Annual Savings	Alternative 1	Alternative 2	Alternative 3
Product	$ 50,000	$ 60,000	$ 40,000
Labor	310,000	310,000	250,000
Power	200,000	250,000	160,000
Chemicals	70,000	120,000	50,000
Total	$630,000	$740,000	$500,000

Evaluation Criteria

The criteria to be used for this example are as follows:

- Annual cost
- Present worth
- Rate of return

The interest rate to be used is 12% and the useful life of the system is 15 years.

Annual Cost Comparison

The annual cost for an alternative is the capital cost of the alternative multiplied by the "Capital Recovery Factor" (CRF) for the interest rate and useful system life. The CRF is calculated from formulas or interest tables available in economics texts [1]. For our example, the CRF is 0.14682 (12%, 15 years). The annual cost for each alternative is, therefore:

Alternative 1 = $2,500,000 × 0.14682 = $367,050
Alternative 2 = $2,850,000 × 0.14682 = $418,437
Alternative 3 = $2,225,000 × 0.14682 = $326,675

Net annual savings for each alternative, are therefore:

Alternative 1 = $630,000 − $367,050 = $262,950
Alternative 2 = $740,000 − $418,437 = $321,563
Alternative 3 = $500,000 − $326,675 = $173,325

Present Worth Comparison

This method requires a calculation of the present worth of future savings. In our example, the yearly savings are assumed to be constant, although in reality they probably will increase at least at the rate of inflation. More sophisticated analysis, which includes prediction of future cost increases, can be employed; again, the reader is referred to an appropriate text.

The present worth of future savings in this example can be calculated by multiplying annual savings by the "Present Worth Factor (PWF)," which, for a uniform series of savings at a 12% interest rate for 15 years, is 6.811. The present worth of savings for the three alternatives is, therefore:

Alternative 1 = $630,000 × 6.811 = $4,290,930
Alternative 2 = $740,000 × 6.811 = $5,040,140
Alternative 3 = $500,000 × 6.811 = $3,405,500

The net present worth of each alternative is, therefore:

Alternative 1 = $4,290,930 − $2,500,000 = $1,790,930
Alternative 2 = $5,040,140 − $2,850,000 = $2,190,140
Alternative 3 = $3,405,500 − $2,225,000 − $1,180,500

Return on Investment

Another method of determining the alternatives of investments is to calculate a return on investment (ROI), which is calculated by dividing capital cost by annual savings. The quotient is equal to the PWF for an equivalent return over the period of interest (15 years, for our example).

For the three alternatives under consideration, the equivalent PWF is as follows:

Alternative 1 = $2,500,000/$630,000 = 3.97
Alternative 2 − $2,850,000/$740,000 = 3.85
Alternative 3 = $2,225,000/$500,000 = 4.45

The equivalent ROIs for these PWFs (derived from appropriate interest tables) are as follows:

160   PROCESS CONTROL COMPUTER SYSTEMS

   Alternative 1 = 24%
   Alternative 2 = 25%
   Alternative 3 = 21%

The calculated PWF factors also can be considered as payback periods in terms of years required to return the initial investment from annual savings.

Each of the economic evaluations for the example case shows that alternative 2 is the most attractive choice, given that sufficient capital is available for the initial investment.

Noneconomic Evaluation

Ranking of alternatives by noneconomic benefits requires utilization of the following evaluation criteria:

Criteria	Weight
Centralized data acquisition	10
Centralized control	10
Centralized data storage and recording	10
Capital cost less than $3 million	10
Implementation less than 2 years	10
System reliability greater than 99%	10
Maintainable by existing personnel	10
Pump malfunction alarms	5
Graphic display capability	6
Trend display capability	7
Additional flowmetering	4
Additional pressure monitoring	8
Link to existing computer system	3
Equipment run-time monitoring	9
Power demand control	9

The first step in the evaluation is to "score" each of the alternatives as to how well each meets the above criteria. Ideally, all would satisfy each of the "10" criteria at the 100% level; however, realistically these may be varying degrees of fit to these "must" criteria.

This part of the analysis is subjective and susceptible to personal bias on the part of the evaluator. One way to minimize the bias problem is to let each member of the task force score each alternative and then average the results.

For the example used in this chapter, each task force member scored the three alternatives in relation to the criteria using a score between 1 and 10, depending on the degree to which the alternative could satisfy the given criteria. The total value for each alternative was the sum of the scores times the weights.

The results of the scoring are shown in Table I. Alternative #2 has the highest score because it fits every criterion to a high degree. The only disadvantages of alternative 2 are a longer implementation time and a requirement to add maintenance personnel. Both disadvantages are related to the higher degree of sophistication provided by alternative 2, which results in greater economic and nontangible benefits.

Note that the economic and noneconomic evaluations will not always favor the selection of the same alternative. In our example, alternative #2 scored the highest in both evaluations. If, however, alternative #1 or #3 had scored higher in the noneconomic evaluation, it would have been necessary to resolve the conflict between the economic and noneconomic analyses. Typically, the economic analysis will be the overriding factor because the noneconomic analysis is more subjective.

Table I. Sample Alternative Evaluation

Criteria	Weight	Alternative #1 Score	Alternative #1 Value	Alternative #2 Score	Alternative #2 Value	Alternative #3 Score	Alternative #3 Value
1	10	10	100	10	100	10	100
2	10	9	90	10	100	9	90
3	10	10	100	10	100	9	90
4	10	10	100	10	100	10	100
5	10	10	100	6	60	10	100
6	10	9	90	9	90	8	80
7	10	8	80	5	50	10	100
8	5	10	50	10	50	10	50
9	6	8	48	10	60	5	30
10	7	8	56	10	70	5	35
11	4	10	40	10	40	10	40
12	8	10	80	10	80	8	64
13	3	2	6	10	30	2	6
14	9	10	90	10	90	5	45
15	9	5	45	10	90	3	27
Total Value			1075		1110		957

Risk Analysis

Evaluation of potential risk is an important factor in any business venture. Factors to be considered include the following:

- State-of-the-art versus proven reliability
- Sophistication versus chance of failure
- Estimated benefits based on theoretical calculations.
- Probability of operator acceptance
- Maintenance requirements versus personnel capability

Evaluation of risk is extremely subjective. The task force must summarize risk factors to upper management as part of the final recommendations. In the example used in this chapter, alternative #1 was the final recommendation of the task force, as well as the choice of upper management. The following was the rationale supporting this decision:

> While alternative #2 appears to present the greatest payoff in both tangible and intangible benefits, we feel that the risk of failure due to complex software requirements is high. This is reflected in a longer implementation time (three years versus two years) and a requirement for additional computer system maintenance personnel. We therefore recommend that alternative #1 be approved for final design and implementation. We further recommend that alternative #1 be designed with the capability to add the advanced optimization software included in alternative #2. In this way, we can install a system with less risk of problems and gain experience that will enable us to implement the optimization programs on an incremental basis.

Regardless what analytical tools are used to evaluate the alternatives, sound management judgment must be applied to make the correct decision. In our example, the analysis showed that while alternative #2 presented the greatest economic and intangible benefits, the associated risk factors were such that alternative #1 was the better choice.

IMPLEMENTATION PLAN

Once the control system is selected, an implementation plan should be prepared. This plan should include the following items:

- Major milestones necessary to design, install and start up the control system

- Schedule based on milestones
- Cash flow requirements related to the schedule
- Manpower requirements

Major Milestones

The following are typical milestones to be included:

Design Phase

- Process and instrument drawings
- Control device specifications
- Electrical drawings and specifications
- Instrument loop drawings
- Panel drawings
- Instrument specifications and schedules
- Input/output (I/O) point list
- Wire lists
- Conduit schedules
- Hardware specifications
- Software specifications
- Report requirements
- Display requirements
- Control strategy specification

Implementation Phase

- Vendor prequalification
- Bid evaluation
- Work statement definition
- Contract finalization
- Hardware submittals
- Software submittals
- Instrument submittals
- Panel submittals
- Control strategy submittals
- Data base generation
- Report generation
- Display generation
- Control strategy generation
- Software integration and test
- Factory tests
- Field installation

- Field startup and testing
- Training
- Final documentation
- Final acceptance
- Process optimization

A brief definition of each design and implementation milestone should be written and included in the plan.

SCHEDULE AND CASH FLOW

A schedule should be prepared that shows the major milestones, time to complete each one and their interrelationships. The final result will be the total project schedule. It may be desirable to combine some of the milestones listed above for clarity, e.g., all submittals could be shown as a single line item on the schedule.

Related to project schedule are cash flow requirements. Projections typically are made on a yearly basis, or possibly on a monthly basis if company needs so dictate. The cash flow schedule may show only disbursements, although it may be desirable to show both disbursements and savings after system implementation. The latter approach provides a graphic illustration of the time required to recover the initial investment.

The method of illustration is optional; however, graphic techniques such as bar charts can be more effective than tabular summations. A cumulative project cash flow schedule for our example, including both disbursements and income, is shown below. This shows that approximately four years are required for the savings to pay back the initial investment following system implementation. Compare Figure 2 to the tabular summation of cash flow (Table II).

Table II. Summation of Cash Flow

Year	Disbursements ($)	Savings ($)	Cumulative Cash Flow ($)
1	1,000,000	—	(1,000,000)
2	1,500,000	—	(2,500,000)
3	—	630,000	(1,870,000)
4	—	630,000	(1,240,000)
5	—	630,000	(610,000)
6	—	630,000	20,000
7	—	630,000	650,000
8	—	630,000	1,280,000
9	—	630,000	1,910,000

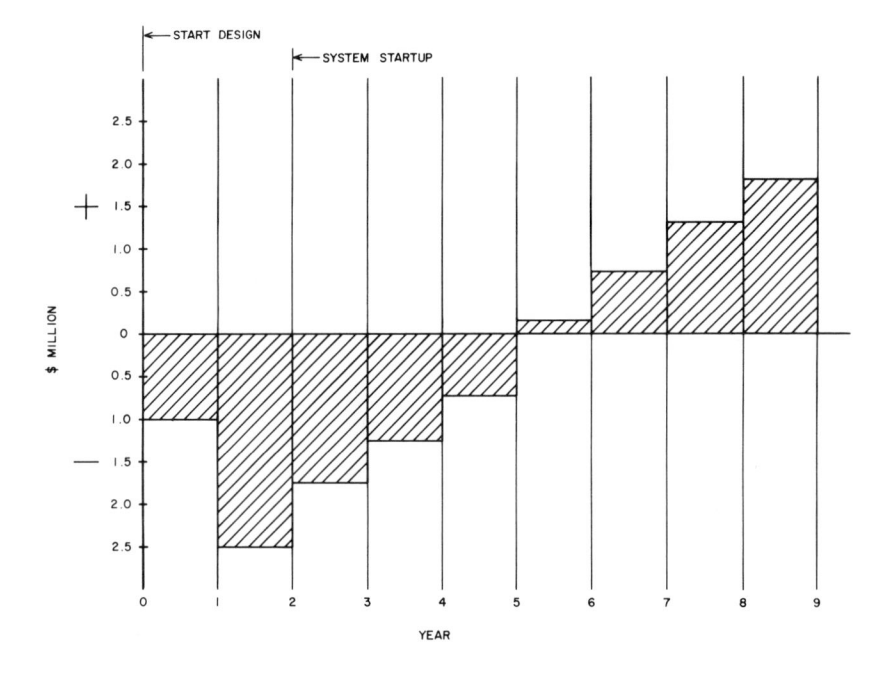

Figure 2. Water system automation, cumulative cash flow.

Note that design and implementation occur during the first two years. The control system does not become operational until the beginning of the third year.

Obviously, the bar chart is a better visual aid than the tabular format. As projects usually have to be "sold" to upper management, good visual aids can be very important in illustrating project attributes.

STAFFING REQUIREMENTS

Operation and maintenance personnel requirements for the project should be identified clearly. If additional personnel are required, salary and fringe benefits associated with those personnel should be included in the project budget. Training for system O&M must be included in the milestone schedule. The implementation plan should include a summary of the type of training to be provided for the new control system.

SUMMARY

Proper planning for control systems is essential to system success. Planning ensures that system objectives are identified early on and that system design can proceed on an orderly and correct basis. Analytical techniques can be used to evaluate alternatives; however, sound technical and managerial judgment must be brought to bear in the final decision, especially when evaluating risk factors.

A good project plan will provide the foundation on which the system can be designed and implemented. The plan will provide guidelines that can be followed throughout the project life cycle. The ultimate success of the control system will be measured by how well it meets the needs and objectives identified in the project planning phase.

Good planning is the first step in ensuring control system success. In summary, the following Do's and Don'ts are recommended during the project planning phase:

Do's

- Do form a project task force.
- Do establish needs and objectives.
- Do formulate evaluation criteria.
- Do identify alternatives.
- Do evaluate alternatives.
- Do select the recommended system.
- Do develop an implementation plan.

Don'ts

- Don't fail to plan.
- Don't fail to write it down.
- Don't overlook any alternatives.
- Don't decide strictly on the basis of price.

REFERENCE

1. Grant, E. L., and W. G. Ireson. *Principles of Engineering Economy* (New York: The Roland Press Company, 1964).

CHAPTER 8

SUCCESS STEP 2 – DETAILED DESIGN

Claude N. Williams, PE

INTRODUCTION

The objective of the design phase of a control system is to expand and define in detail the ideas, goals and concepts determined during the planning phase. The result of the design phase is a set of specifications that define exactly what will be built. These may be used as a contract document to buy a control system or used as an in-house reference. In either case it must be clear. It must have also enough detail to ensure that the system is built right. An unclear or incomplete specification is the biggest source of problems in a control system.

A computer-based control system has several components common to analog instrument systems, including control panels, instruments and manipulated devices. Other components are unique. The computer-based system has attributes that can affect maintenance, laboratory, inventory and administrative procedures, as well as process control. To achieve maximum benefit it is extremely important for those people who will use the system to be involved in the design phase.

The purpose of this chapter is to explain how the small details and user desires are combined to define a control system that will be successful. A successful system can be defined and attained.

Top-down design is the best philosophy for control systems. It allows effective organization of the user, design team and vendor. Given user inputs, the design team prepares the specification. This is the blueprint against which the vendor adds the low-level detailed design.

DESIGN HIERARCHY

Any design project must be organized to be effective. What is supposed to be done first should be. Subordinate decisions, such as component details, can be made later. Revisions also must not be handled haphazardly. Revisions made in design tend to affect a wider range of areas than usually expected, so the organization must allow for change and for assessing and limiting the effects of the change.

A proven concept that has such organization is "top-down" design. In top-down design the system is designed in levels, each composed of one or more modules. Design is started with the top level, in which the system as a "module" is considered. Its goals, attributes/limits, needs and performance are defined. In turn, the system can be divided into subsystems, which are defined in the same manner. This process continues until design of the lowest-level modules has been completed. As design changes occur, they can be dealt with on a modular basis. Even if more than one module is involved with a change, the repercussions usually are confined to a few modules, leaving the rest of the design intact. An example of top-down design is shown in Figure 1.

Using this hierarchical method, the scope of the design task can be determined fairly well by answering two questions:

1. What needs to be designed?
2. What amount of detail is needed?

Once these are resolved, design is ready to proceed from the planning phase of the project.

One result of the planning stage is a rough idea of the system and perhaps bits and pieces of other modules. Any input from the planning stage should be reevaluated at the start of design, particularly if a major time gap exists between these two steps. The need for reevaluation is not a reflection on the quality of planning but rather on the fast-changing technology and the changing needs of the user. This will occur throughout design, implementation and the life of the system.

Usually this poses no problem. Much effort is needed to freeze ideas long enough to finish design and implementation. For this reason, it is desirable to establish "freeze dates" for each module, after which it cannot be changed. In practice, freeze dates often melt. When this happens, several nontechnical constraints such as budgets and deadlines come into play that may lessen the design quality. Situations such as this test the soundness of the design structure and the quality of the design team. The remainder of this chapter develops the top-down design philosophy.

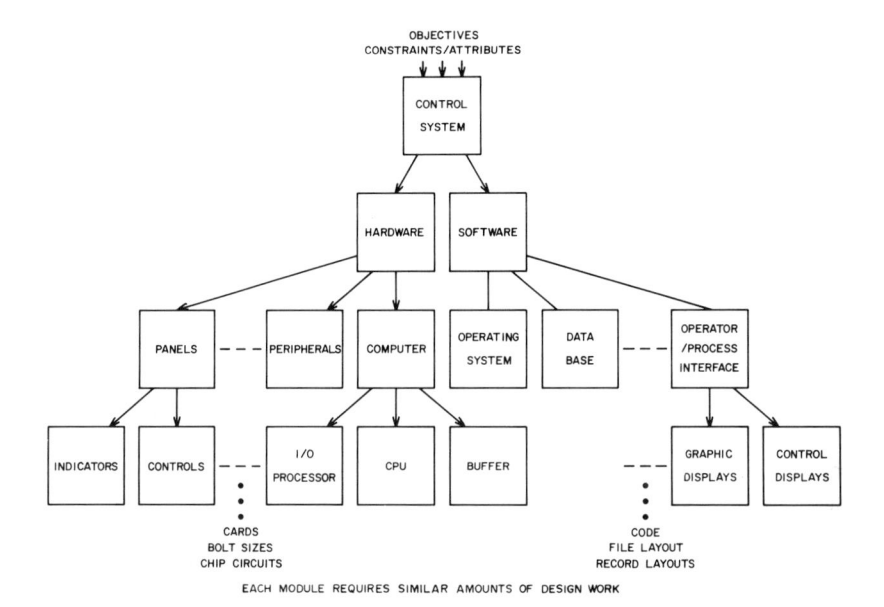

```
                    OBJECTIVES
               CONSTRAINTS/ATTRIBUTES
                    ↓  ↓  ↓
                  ┌─────────┐
                  │ CONTROL │
                  │ SYSTEM  │
                  └─────────┘
                   ↙       ↘
            ┌──────────┐  ┌──────────┐
            │ HARDWARE │  │ SOFTWARE │
            └──────────┘  └──────────┘
```

Figure 1. Top-down design.

DECIDING WHAT TO DESIGN

One question posed in the previous section was, "What needs to be designed?" A quick and correct answer would be, "everything." However, if the question were rephrased to "What do *I* need to design?" the answer may not be so simple. It would depend on who is asking the question. The answers may be restricted to four possible persons or teams:

1. **User**—the one who uses or operates the control system.
2. **Engineer**—the one who does prepurchase design work. It might be an in-house engineering department or an outside consulting firm.
3. **Vendor**—the one who supplies the hardware and software being purchased. There might be more than one vendor involved.
4. **Owner**—the one who is purchasing the control system. Except for rare cases of generosity, the user is part of the owner. It is also possible for the engineer and, in very large corporations, the vendor, to be part of the owner.

As the owner could be in any of the other roles, its design duties will be made up of the others. Discussion of this composite person is therefore omitted. A general relationship of these persons is shown in Figure 2.

User Design

Too often the user is not involved enough in design; in some cases, not at all. This is a serious mistake. Vital information on what the system should be able to do and not do will be missing. The result will be something less than desired and something that will not be accepted. In fact, excluding the user from design will lead to poor feelings and less than full use of the system by the user. While feelings, usage and acceptance may not be classed as technical design criteria, they are very important to the success of the control system.

On the other hand, users often are not qualified to do detailed design, whether at the board level or the system level. What is really needed from the users are system goals and attributes of such modules as process control, operator/process interface (OPI) and data reporting. These are areas that the user will have to interact with on a daily basis. Customizing the design to fit the user at these points requires user input. This will help make the system acceptable once it is ready. User design is shown in Case Study I.

Case Study I: User Design

Addition of a large oil/coal-fired boiler was being planned for Joe's Pulp Mill. The unit was to be digitally controlled with a large number of control loops and interlocks. Joe, the mill manager, was concerned about the startup of such a complex unit. His design input for boiler control was as follows:

1. Startup will be by one person in less than one shift.
2. Startup procedures are needed for winter starts, hot starts and cold starts.
3. The operator should be able to abort the starting sequence at any time.

Note that these goals and limits help define the system, control software and operator/process interface.

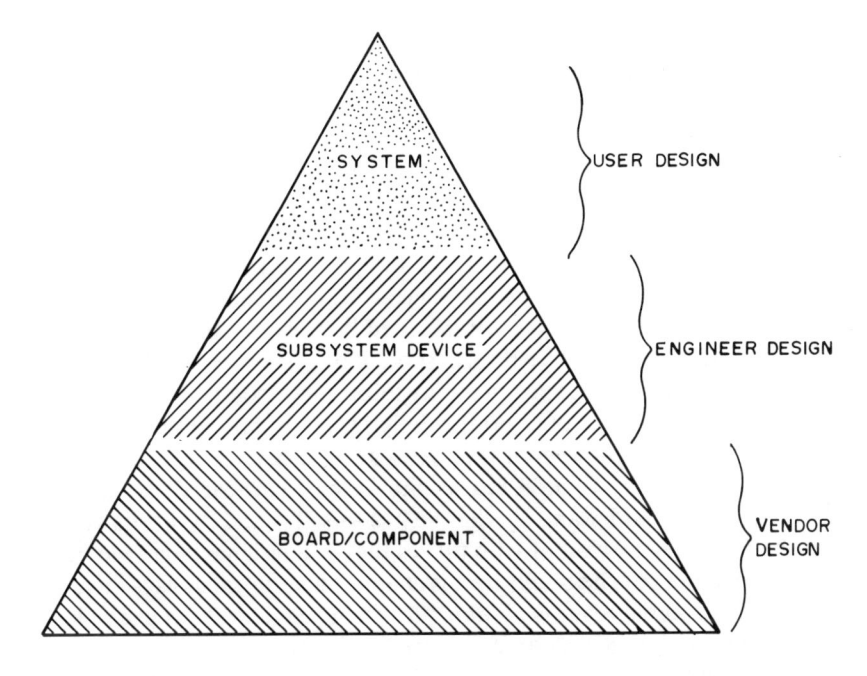

DESIGN AT DIFFERENT LEVELS IS USUALLY DONE BY DIFFERENT PERSONS

Figure 2. Personnel involved in the design.

Engineering Design

Engineering design is the act of turning user requirements into a set of written specifications. These, in turn, become the requirements of the vendor's design. This definition of engineering design closely resembles the definition of system design given in the introduction. In fact, "design" most often means this particular aspect of the design process. Unless otherwise stated, "design" will refer to engineering design.

Control system design usually is done by a team composed of specialists in systems engineering, programming, operator interface, process control, instrumentation and hardware. The team and its manager first will rough out the system design and then split up, with each person performing the top-down design in his area or areas. The scope of each person's work depends on the size of the project and the level of detail required. It does not depend on whether the specifications are used in-house or for outside contract. Management of the team and coordination of the control system design with mechanical, electrical and structural designs are very important tasks. They will be discussed later as an aspect of the control system design guide.

Design usually defines device-level modules for hardware or program-level modules for software. For example, design of video display terminals (VDT) will include the number of units needed, their location, environment and use, color versus black and white, screen size, resolution, adjustments (vertical hold, brightness and power on/off), health, safety, mounting type, communications interface and special features. Board (component)-level details are not included in "design," but will be determined by the vendor. The design should be general enough to allow several vendors to meet specifications to promote competitive prices.

The choice of who should do the design work is among the user, consultant or vendor. The vendor may seem an attractive choice because the costs are free (really figured into capital costs). However, the design is suspect; "Is it what I want or what he has to sell?" and "Will the system cost be fair?" Discussion of who makes the best design engineer appears in other chapters of the book. In this chapter, an unbiased, well-meaning and knowledgeable team is assumed to be the engineer. An example of engineering design is presented in Case Study II.

Case Study II: Engineer Design

Joe, the mill manager in the previous case study, contacted the Dictator Control Co., which would do the engineering and also could furnish the complete control system. Dictator's design solution to the startup problem was as follows:

1. A VDT and a log printer would be located in the boiler control room with the process computer.
2. The VDT would show the operator current operating data and allow the operator to make control changes including startup abort.
3. The printer would generate a copy of whatever operating instructions the operator requested.
4. Between the ability to enter commands via the VDT and the printed instructions, it was suggested that an experienced operator could do startup in one shift.

Joe did not like the price of Dictator's system and was not sure about one shift startup, so he hired Abby's Consulting to do the engineering. Abby prepared a set of specifications for the following system:

SUCCESS STEP 2—DETAILED DESIGN 173

1. A VDT would be put in the boiler control room with the process computer. Data reporting would be done by linking to the mill's central computer.
2. The VDT would show the operator current process data and allow the operator to make control changes, including startup abort.
3. A program would be written for computer-assisted startup. The program would lead the operator through the startup sequences and do about half the startup work by interactive data entry (IDE) on the VDT. Estimated time for startup was two to four hours.

Joe's input affected such things as peripheral choice and control programs. Even so, the engineering design took two different directions. The two systems would cost about the same, with hardware costs in system A offset by higher software costs in B. Yet system B has a better chance of success. It adds to the mill's central reports; it uses a powerful tool—IDE; and Joe can identify this use of IDE as a result of his design input.

Vendor Design

When the control system vendor begins to implement the engineering design, a major degree of low-level design (component/board/record file) may be involved. The actual amount on a given job depends on the vendor and how much standard design can be used. The technology is fast-changing so that a portion of the vendor's standard items will be obsolete, thus requiring some new low-level design in any case. Generally, a top-down design or a lateral design (modification of existing designs) approach is used. An example is given in Case Study III.

Case Study III: Vendor Design

Dictator Controls won the job of furnishing the equipment and software described in system B of the previous case study. The remainder of the system design was accounted for in the following manner:

1. The VDT was purchased by Dictator from another vendor. A desktop, stand-alone model was chosen per the engineer's specification.
2. The process computer and other hardware used the same design as in previous jobs done by Dictator.

3. The IDE program was designed top-down based on the criteria in the engineer's specifications.
4. Other software used the same design as in prior jobs by Dictator. Some improvements were made.

In this example, new vendor design was limited to a small portion of the system, which is often true in practice. It is also often true for new vendor design to be spent upgrading "standard" vendor products.

User/Engineer/Vendor Relationships

Three levels of design—system, device/subsystem and component/board—have been given names based on who does the design work: user, engineer and vendor, respectively. Who does what may vary, but it is convenient to categorize design in this manner. Another relationship that clarifies the roles of the persons involved is the buyer-seller relationship. A person buying a control system should not do so unless there is adequate assurance that the system is needed and can meet the needs. The user is clearly the buyer and the vendor is the seller, but on which side is the engineer? Usually the engineer provides the assurance needed by the buyer and is thus placed with the user. Often user and engineer are thought of as the owner, which means the buyer versus seller relationship is an owner (user and engineer) versus vendor relationship.

HOW MUCH DESIGN IS ENOUGH?

The second question posed to set the scope of design work was, "What amount of detail is needed?" As before, the answer depends on who asks the question: user, engineer or vendor. The user is concerned with defining system objectives, the engineer with transforming these into specific device functions, and the vendor with generating the fabrication plans.

This does not mean that design is vague at the top level and clear at the bottom level. The top-down concept breaks the design into modules of roughly equal work. It will only produce the desired result if each module is thoroughly and completely defined. In developing a given module, the designer takes its goals, attributes/limits and performance needs and generates detailed plans. The detailed plans of one level become the goals, attributes/limits and performance needs of the next lower level. This pyramid arrangement is analogous to a corporate personnel structure.

SUCCESS STEP 2—DETAILED DESIGN

The president provides the highest-level direction, and the lowest-level employees produce the product. Unclear direction anywhere in the corporate ladder will reduce efficiency and, in the end, profits, which is the corporation's measure of success. In the same way, unclear work anywhere in the design process will result in a system that does not meet expectations. One way to view this is shown in Figure 3.

This book is not intended to present an in-depth review of component-level design. Therefore, vendor design will not be expanded on further, although the topic will arise from time to time. To avoid awkward cross-references, user and engineer design will be grouped together as "designer." Keeping in mind that design must be thorough and complete, the question "How much design?" still needs more of an answer.

Design Options

There are some choices available to the designer that can influence the amount of detail needed to define a device or program. One choice is whether a standard "off-the-shelf" item can be used or whether it must be a special application. Standard items that fulfill system needs have several advantages:

1. Less design work is needed. Specifying a comparatively small number of parameters and a standard of quality will result in the desired item, which already will have shown itself to work.
2. Item costs will be less. New design by the vendor can be avoided, which means that the cost will be lower than for a special product.
3. System support will be better. Maintenance, obtaining spare parts and troubleshooting will be easier.

Most hardware and some software can be designed as standard items. Hardware devices that are usually not handled this way are operator consoles and sample systems for instruments. Consoles are heavily influenced by aesthetic factors and project needs. When standard items are not adequate, the design of custom items is required.

Another choice is whether to buy everything or to do some of it oneself. The best candidate for do-it-yourself work is process control software. User needs in this area change rapidly. Change often occurs in the time between buying the equipment and startup of the control system. Rework costs could be avoided by not having the vendor provide control programs. Problems encountered by doing this are presented later.

A third choice is whether to obtain the control system via a direct bid

176   PROCESS CONTROL COMPUTER SYSTEMS

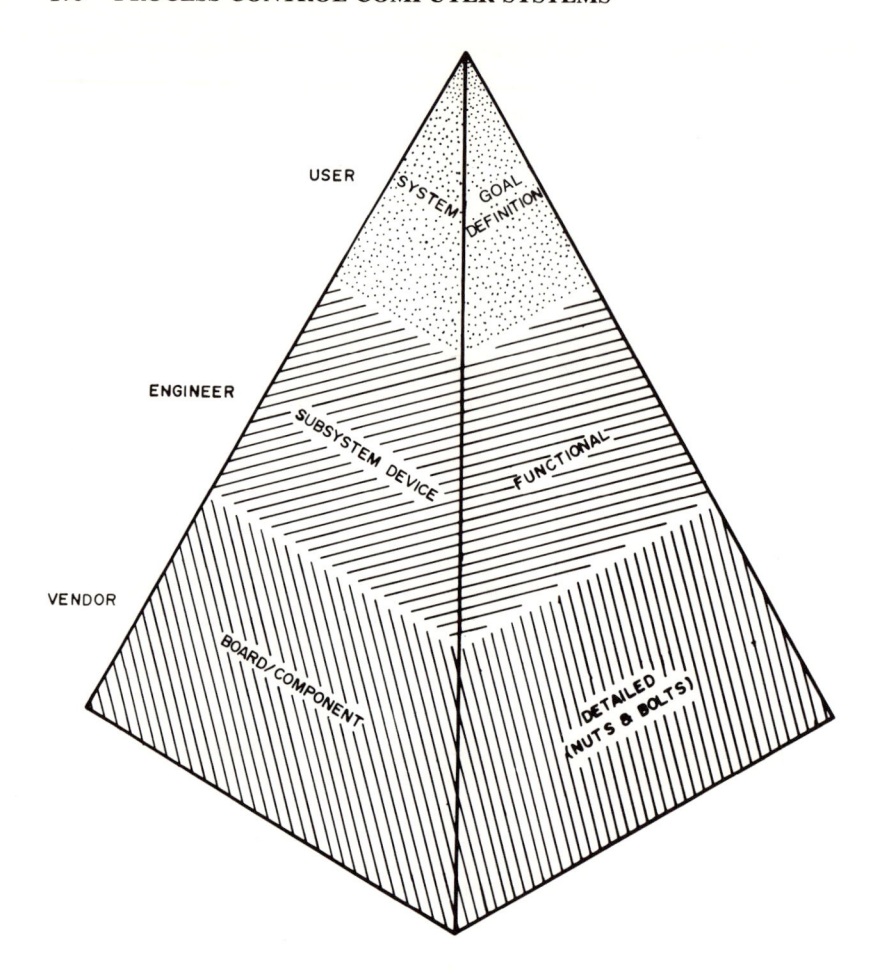

Figure 3.   Design pyramid: shows how much design is necessary.

or through the services of a general contractor as part of a larger contract. Direct bids offer more freedom to work with the vendor; however, the vendor's responsibilities and coordination with others need to be more definitive than for indirect bids.

A fourth option is whether to define the items functionally or explicitly. Explicit specifications state how a task is to be done. Functional designs state what the task is. For example, to say that a keyboard must be centered on and abut a VDT is explicit. To require that a keyboard be provided for easy use while viewing the VDT is functional. Functional design allows the vendor to use its design experience to provide the desired functions. However, care must be taken to compel the vendor to do good work; generally, functional designs require less design work. Thorough and complete design is still necessary. An example of using design options is shown in Case Study IV.

Another factor that determines the amount of design is the timing of the design work. For maximum benefit, the control system should be designed with the process piping and equipment. Early analysis of process control will produce a better system. It will improve piping arrangements, pipe sizes, valve selections, valve sizes and equipment design. It also will make for uniform application of instrumentation and panels and for adequate interface to the control system. In the early stages, control strategies should be written in a preliminary form. This will guide instrument placement as shown on preliminary process and instrumentation diagrams (P&ID) and instrument selection.

Case Study IV: Design Options

1. Standard versus special.
2. Design now versus do-it-yourself later.
3. Direct bid versus subcontract.
4. Functional versus explicit.

The following specification for a control strategy is part of a control system contract to be awarded to the lowest-qualified bidder.

> Acid is to be added to a process stream and mixed in a small tank. A software controller, pHC, shall modulate a control valve on the acid feed line to keep the tank effluent pH at an operator-entered setpoint. The controller shall be a direct acting proportional integral (PI) controller having the equations and options specified elsewhere for PI controllers or equivalent equations and options. In addition, pHC shall have two complete sets of tuning constants, one to be used within a pH range about the setpoint and another outside this range.

It was anticipated that pH control might be difficult for this process, which would require additional software refinements. Rather than implement this part of the software, it was decided to defer programming until the control system and process were installed and operating. Provision for future software additions by the owner was included in the specifications:

> The vendor shall provide spare software capacity for a pH conversion module (CONV). The module shall use a second-order polynomial and a 12 × 30 lookup table for one of the constants in the polynomial expression. This spare capacity shall be grouped with the software actually implemented and shall be dedicated software. General provisions for spare capacity shall apply to the total dedicated software.

Large Systems versus Small

Design of a control system for a large processing plant can appear to be a formidable task. Man-years of effort, specialized team members, sophisticated interrelationships between components, new and poorly understood technology all contribute to a mystique of difficulty. As a result, large engineering costs can easily be justified, even for the wrong reasons.

Design of small control systems, on the other hand, tends to be overlooked or, at best, incomplete. It is not valued because of simplicity, few goals, few pieces and easily understood benefits. Yet a complete top-down design is needed. The design pyramid may be small, but the need for good, clear design definitely still exists.

Engineering design for a large control system will cost 10–25% of the installed controlled system price. For small control systems, design costs may exceed system costs. The large relative cost for design of small systems is a major reason for a close look at the benefits of design. Complete design is recommended for the following reasons:

1. Lack of design is a major source of system problems. It may not work at all. In the worst cases, a system may be abandoned.
2. Incorrect design can lead to hidden costs for operation, maintenance and correction.
3. Design omissions will lead to frustration and failure due to missed schedules, misunderstood requirements and unsatisfied objectives.

Rather than omit design, it is better to find a way to lower the costs. Perhaps the cost could be amortized over several similar control systems. Judicious use of design options could lower design cost. Due to small design teams and small capital expenditures, significant savings also can be realized from less structured project communication, reduced design guide and briefer specifications. An example of small system design is presented in Case Study V.

Case Study V: Small Systems

A publishing firm wanted to upgrade the safety interlock system on each of five similar presses. One programmable logic controller (PLC) for each press was to replace the relay logic used previously. The engineer worked with the user to develop a rating method using weighted criteria. The rating method was used to select the programmable controller; programming was done jointly by the engineer and user after installing the equipment. The following were some areas researched during the engineering design:

- Ability and likelihood of future replacement of control loops for tension control and automatic registering
- Compatability with company standards
- Expansion capability
- Centralized data logging
- Operator override capability
- Programming ease
- Correlation with quality control data logging
- Troubleshooting
- Alarm annunciation – diagnostics
- Maintenance
- Installation ease
- How much programming flexibility would be enough

Design costs would have been the same for one press. Spreading the costs over five units and doing the programming at a later date kept the design costs within budget.

THE DESIGN GUIDE

The first step in the design process is to prepare guidelines for members of the design team. The objective is twofold; (1) to keep the detailed work in line with concepts of the planning phase, and (2) to maintain design standards. The material in a guide may vary depending on the design team and on special concerns. All guides will have important information describing:

- Configuration
- Control and monitoring
- Electrical and mechanical coordination
- Documentation and nomenclature
- Project management

Such a guide is indispensable to successful design.

Configuration

One result of most planning stages is the choice of a system configuration. It may be a combination of central computers, distributed computers, distributed multiplexers, central information-interface equipment (VDTs, keyboards, printers, tape units and video copiers) and distributed information interface equipment. The configuration now must be reexamined. The needs of the user should be rechecked. Based on these efforts, the guide should state the user goals and should go as far as possible to define configuration. The following items should be addressed:

- Cost
- Minimizing field wiring
- Relating distributed units to processes
- Acceptable reliability
- Acceptable availability
- Locating sites of all equipment
- Selecting distributed and central information-interface equipment to match staffing plans
- Allowing for future equipment expansions
- Identifying site equipment needs such as heating, ventilation and air conditioning, special air filters and control room facilities

The configuration defined by the design guide is tentative. Changes will be made during the rest of design; however, the effort of preparing the guide is very worthwhile because even with modular top-down designs the amount of rework to make changes later on increases greatly. Small changes in configuration may cause changes in many other areas of design.

The configuration is defined on a drawing called the block diagram. An example is shown in Figure 4. It depicts a system that has mirror image redundancy. The two central computers and dual-ported peripherals make for a redundant, highly reliable system. This configuration reflects the needs of a particular user.

SUCCESS STEP 2—DETAILED DESIGN   181

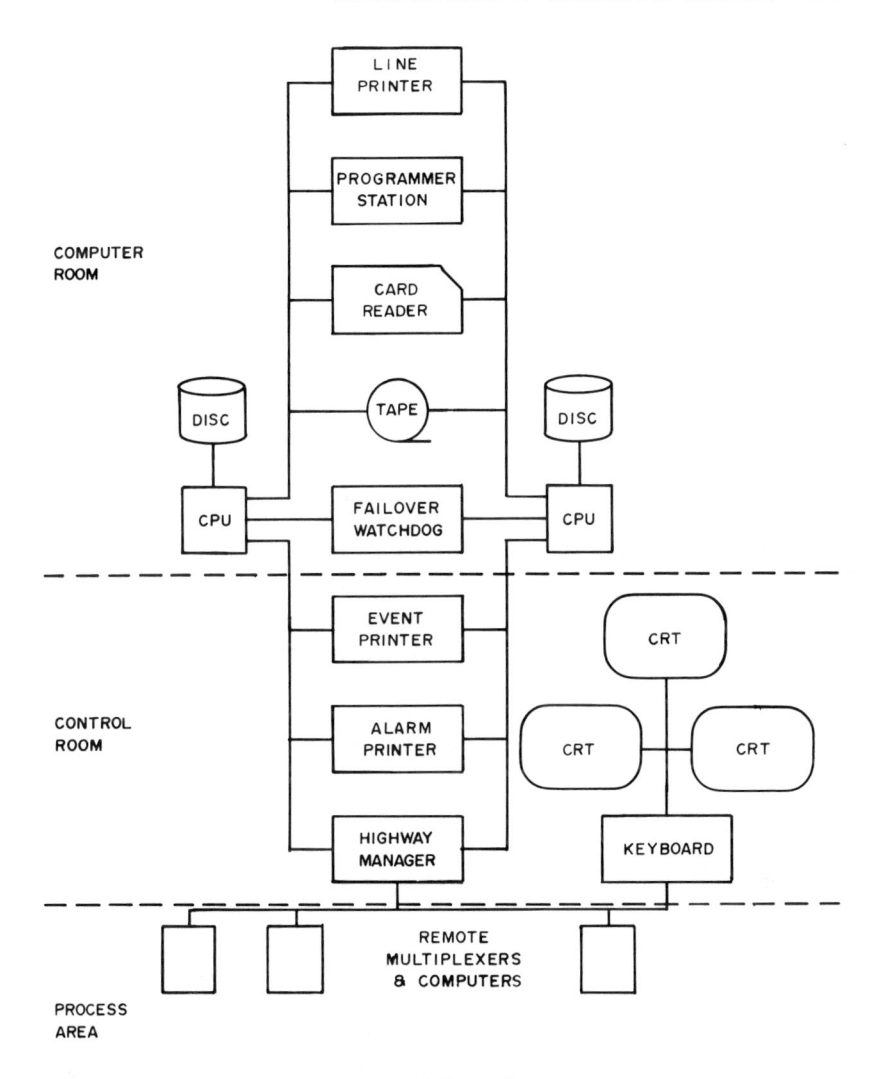

Figure 4.   Block diagram.

Control and Monitoring

During the planning stage, several process control strategies that were required from a cost/benefit point of view will have been outlined. These strategies may have been the basis for justifying a digital control system. If so, the entire plant would need to be reviewed to identify other strategies, which, on their own merit, may be minor, but are very attractive as an additional increment to the control system. The design guide should contain a description of previously outlined strategies and criteria for selecting other control strategies. It also should define what levels of control are to be provided. Is analog, automatic or manual backup to be provided? Is computer-assisted startup or shutdown desired? Will operations be centralized or dispersed?

Electrical and Mechanical Coordination

Team members typically will be given tasks that fall into their special areas of competence. Lack of interaction between these areas leads to compartmentalized design, with big chances for major discrepancies or omissions. For instance, an electrical engineer assigned power distribution may conflict with or inadequately support another electrical engineer assigned instrumentation. Which one shows conduit routing on plans? Which specifies power monitoring instrumentation? Which one interfaces with mechanical equipment? To prevent confusion, the breakdown of responsibility should be defined. Also, points of interfacing should be resolved, such as motor start/stop circuits, valve open/close circuits and field wiring conventions. Chance of confusion also is reduced by establishing common guidelines for electromechanical equipment. Such guidelines often include color coding for wires, panel-VDT indicators and equipment, and panel placement requirements. Guides for reviewing each other's work also should be planned.

Documentation and Nomenclature

Standards for drawing preparation, from amount of detail to line sizes, and standards for documents from content to outline may be included in the design guide. Guidelines for numbering schemes and names are especially important. Much confusion can be prevented by early agreement in these matters. Many numbering and naming schemes may be used on a project—some just for construction and some that last for the life of the equipment. Without coordination, a partial list of designations for a single equipment item might look like this:

- Tank A2-A1   Structural design engineer
- Reactor 14.51D   Contractor
- Train A2, Stage 1   Electrical design engineer
- SO21459-01   Maintenance inventory number
- SAD109   Instrument tag number
- Train B, Reactor 1   Operations label

Project Management

Management is the effective use of personnel and resources to complete design within a given time frame. Management is not more important for control systems than for other systems, nor is it less important. The duties of team members, milestones and deadlines should be set early in the project by using the design guide.

DESIGN OF CONTROLS

The design of a control strategy usually starts with an in-depth review of the process to be controlled. As ideas arise, they are put onto a schematic of the process called a process and instrumentation diagram. Development of the strategy continues using other diagrams, flowcharts and notes. Eventually, a complete control concept evolves. It defines what control is performed, where it is performed, what backup is needed, what instrumentation is needed and what control equipment is necessary. These steps for control strategy design are listed in Table I and are described in more detail in the following paragraphs.

Controllability

The process and equipment to be controlled are first analyzed to determine whether the desired result can be achieved. For example, splitting flows between two outlets of a liquid storage tank by using fixed weirs is uncontrollable. That is, there is no means by which an operator could change the flow split. Due to inexactness of weir construction and variable hydraulic conditions, the flow split also will not be according to the design intent. To control flow splits, valves and a level transmitter might be used. The valves allow the flow split to be changed (i.e., controllable), and the level transmitter provides an operator with information on whether the valves need opening (high liquid level) or closing (low liquid level). The valves could be operated automatically if a level controller were added to complete the loop. This control concept can be shown in diagrammatic form, as in Figure 5.

Loop Configuration

A control loop is comprised of three essential parts: sensor, controller and end device. In this case, the sensor is a level transmitter that measures

184   PROCESS CONTROL COMPUTER SYSTEMS

Table I.   **Control Strategy Design Checklist**

A. Controllability
   1. Can the desired result be achieved?
   2. Can the process be controlled?

B. Loop Configuration
   1. Propose loops and select the best one.
   2. Prepare control diagram.

C. Process and Instrumentation Diagram
   1. Prepare P&ID.
   2. Maintain a data base.

D. Algorithms
   1. Add contingencies.
   2. Provide manual control.
   3. Define strategy startup.
   4. Define strategy shutdown.
   5. Prepare flowchart.

E. Analysis and System Design
   1. Check stability.
   2. Make simulation.

F. Instrumentation
   1. Choose sensor—make head loss calculations for inline sensors.
   2. Define transmitter capabilities.
   3. Design sample handling.
   4. Choose and size control device.
   5. Plan signal handling.
   6. Consider maintenance, mean time between failures (MTBF) and mean time to repair (MTTR).
   7. Design mounting and installation.

a parameter, level, called the process variable. The controller compares the process variable with a setpoint and determines what control action is needed. The controller's output moves the end device, in this instance two valves, to reduce the difference or error between setpoint and process variable. For this loop to be useful to control flow split, two improvements must be made.

First, some way is needed to set the flow to any desired flow split, not just even. This can be done by multiplying the signal to each valve by a fraction. If the fractions are set properly, a flow split of 1/2–1/2,

1/4–3/4 or any other desired split can be obtained. The second improvement is to measure the flow in each pipe. Until now, the control strategy assumed that the flows were proportional in some way to the valve position. Nonlinearity in valve position/flow characteristics and hydraulic differences may cause significant errors. Measuring the controlled variable

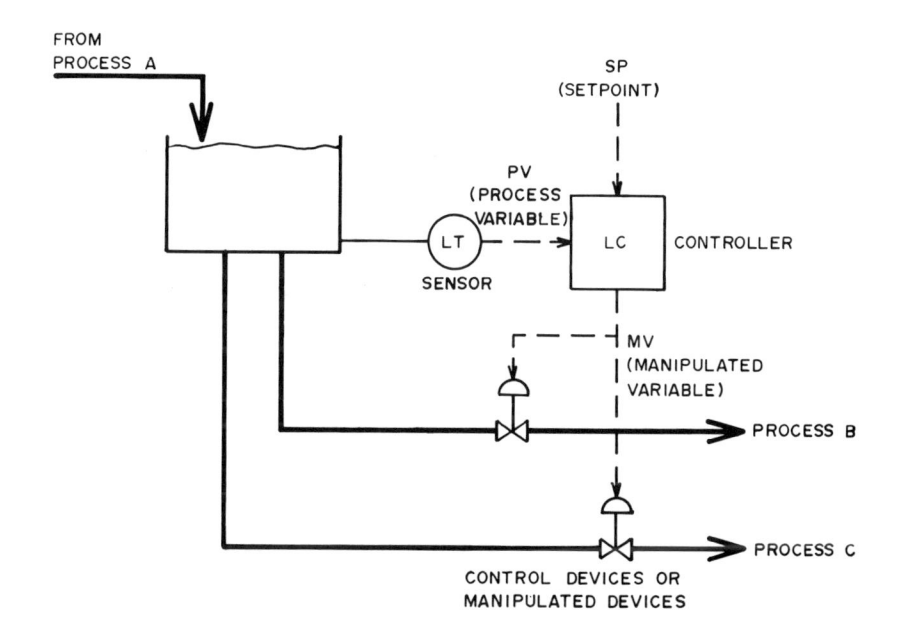

Figure 5. Simple control diagram.

directly instead of by inference removes this source of error. To close the loop with these two new process variables, two additional controllers must be added in what is called a cascade arrangement, in which the output of one controller, the level controller, becomes the setpoint of another, the two flow controllers. The two improvements are shown in Figure 6.

At this point, the control strategy is complete; all the desired functions are present and the method is not too complex. It should be considered tentative until the freeze date for control strategies is reached. Further work now should be done to define where the strategy is to be performed. It can be done either by analog controls or by software or any combination with or without backup. The choice depends on three factors: performance standards, control philosophy and control system configuration. Performance standards would include the following:

186 PROCESS CONTROL COMPUTER SYSTEMS

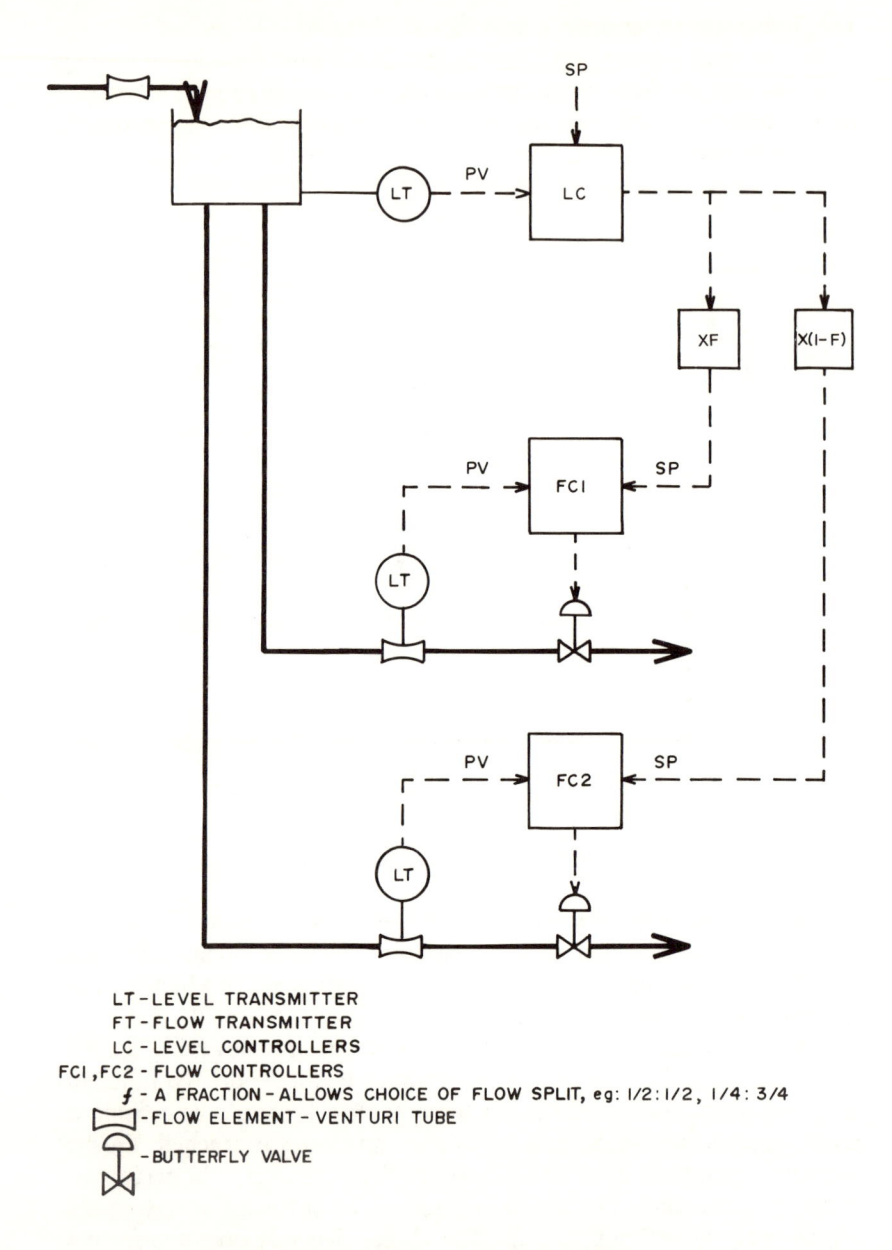

LT - LEVEL TRANSMITTER
FT - FLOW TRANSMITTER
LC - LEVEL CONTROLLERS
FC1, FC2 - FLOW CONTROLLERS
f - A FRACTION - ALLOWS CHOICE OF FLOW SPLIT, eg: 1/2:1/2, 1/4: 3/4
- FLOW ELEMENT - VENTURI TUBE
- BUTTERFLY VALVE

Figure 6. Control diagram — cascade control.

Performance Standards

- **Availability**—the percentage of time that the system is ready to perform control. Availability is also applicable to each component of the system.
- **Reliability**—probability of successful operation of a device.
- **Response Time**—the time a system takes to react to an input signal or command. Examples are the time to show a requested display or the time to shut down a pump on receiving an alarm.
- **Contingencies**—the desired performance in case of failures.

Performance standards must be interpreted with respect to the configuration. For example, two controllers wired in parallel to each other have a higher availability than either controller alone. In contrast, two controllers wired in series will have a lower availability than either alone.

Let us continue with the flow split example. Assume the availability of the computer or programmable controller is 99% and the availability of an automatic hardware controller is 99%. This means that either location of the strategy will result in 90 hours per year when no automatic control will be performed. While the automatic controls are inoperative, a person must be present to manually open and close the valves. As either a software or a hardware location alone results in a lot of downtime, an arrangement was chosen in which the software control strategy is backed up by a parallel analog controller. The parallel arrangement increased availability to 99.9% or only about 9 hours when an operator is required. The location of controls is shown on a P&ID, a sample of which is shown in Figure 7. Another way to increase availability is to use higher-quality equipment. Many computers, as well as hardware controllers, have availabilities over 99.9%.

Control Philosophy

Philosophy of control defines what type of control is done in the system. There are two types of digital control:

1. Supervisory control and data acquisition (SCADA)
2. Direct digital control (DDC)
   a. Digitally emulated analog control (DEAC)
   b. Optimizing digital control (ODC)

In the example, DDC is the control type chosen to increase the availability of the system. Within DDC, one may select either ODC or DEAC. DEAC, computer duplication of analog controls, achieves three desirable

188   PROCESS CONTROL COMPUTER SYSTEMS

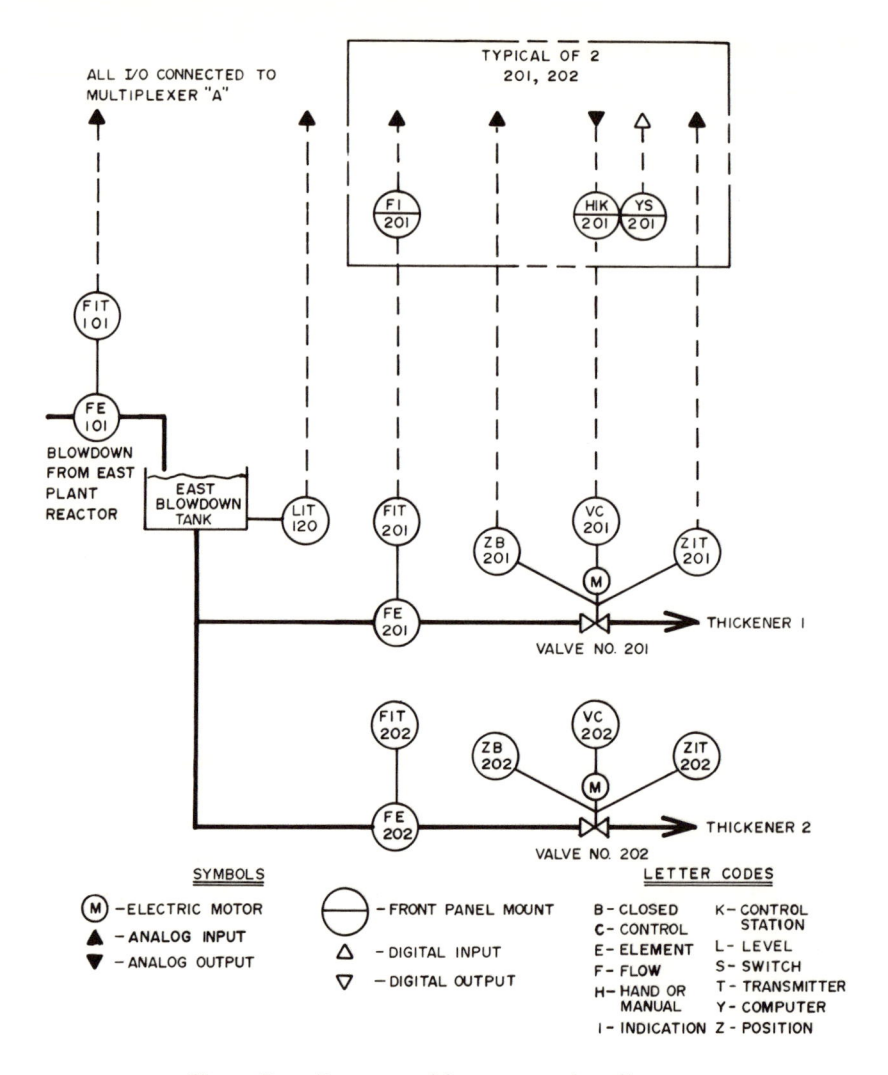

Figure 7.   Process and instrumentation diagram.

objectives: relatively simple control strategies, centralized control and controls already familiar to the operator. These are all worthwhile, but additional benefits potentially can be obtained from ODC, optimizing digital control. This case is seen in Figure 7, in which flow controllers are present only in software. ODC control also would cover enhancements to this strategy. One enhancement might be to use the tank's storage capacity as a buffer to equalize flow over time. This would make the downstream process more stable and thus save energy.

Process and Instrumentation Diagram

The purpose of a P&ID is to show instrumentation and control for a process. It serves as a communication link among members of the design team. Electrical, mechanical, process, instrumentation and control plans can be combined on one diagram. It is the only place where this is done. Although mostly used for instrumentation and control (I&C), the diagram enables the other team members to do their work and provide necessary support.

The contents of the P&ID may take two general forms. One is that all I&C must be included on the diagram from pressure relief valves to control system interface. The other position is to show only I&C directly connected to the digital control system. Where practical, it is best to show everything because then it is complete. Items are less likely to be omitted or duplicated. Coordination is better when using one drawing because the relationship of all the parts can be seen. Often this approach is not practical. Time and money are genuine constraints. If the P&ID is to be restricted, the nature of the restriction should be made clear.

Besides hardware, instruments and analog controls, it is also possible to show software control strategies on the P&ID. Logistics make this difficult to achieve. Even moderately complex processes tend to fill a drawing with I&C hardware, leaving little or no room for software. Software itself could take a full drawing or more to describe. Aside from space considerations, there is the drafting rework required to alter software during design, construction and operation. Software flexibility is a major advantage of digital control systems, but drafted drawings are not as easily changed. A better approach is to put only hardware on the P&ID and document software elsewhere via control diagrams, text or flowcharts.

Algorithms

An algorithm is a method used to arrive at a solution from a set of inputs. More generally, an algorithm is synonymous with control strategy as discussed above. Algorithms are defined by using control diagrams, text or flowcharts. The choice of definition tools depends on project documentation standards. Control diagram and text is the most used combination. Strengths and weaknesses of these methods are as follows:

- **Control Diagram** – excellent for showing loops, poor for showing sequences or nonsimple interlocks.
- **Text** – essential for understanding control diagrams, good for defining calculations.

- **Flowcharts**—good for explaining sequences and interlock logic, poor for showing control loops or calculations and becomes unwieldly for complex logic.

Explanation and examples of control diagrams (Figure 6) were presented in describing the flow split problem. As a comparison, Figure 8 shows the flow split algorithm in flowchart form. Several modified flowcharting methods are also in use. These retain the yes/no binary logic but reduce the need for connecting lines to overcome the tangle of lines in complex cases.

So far, the algorithm has related only to automatic control directed by the computer. Another aspect of algorithm design is to provide robustness. A "robust strategy" will contain contingency actions and allow degraded control to make it less susceptible to total failure. In analog control systems, a component failure often will completely disable the controls, but in digital systems only partial loss should occur. There are three common means for providing degraded control for the flow splitting case:

1. Replacing bad sensor data with operator-entered data is one means. If the tank is large in comparison to influent flow fluctuations, the level control loop will be slow. The operator would have time to enter tank level for a bad level sensor/transmitter without much loss in control efficiency.
2. A second means is manual entry of setpoints to inner loop controllers. Loss of the level sensor/transmitter also could be compensated for by manually inserting setpoints to the flow controllers. This capability also would provide flexibility in normal situations by allowing one constant flow pipe and one swing flow pipe.
3. Manual positioning of valves is a third means. If a flow sensor/transmitter fails, the affected loop should be placed into a manual state where the valve is controlled to an operator-selected position. Approximately equal flow could be maintained by selecting the same position as the valve remaining in the cascade loop.

Contingency actions are automatic digitally directed responses to failures or alarms. An example of a situation requiring contingency action can be shown in the flow splitting example. Suppose that the tank were small and fed by either or both of two pumps. Normal changes in pump speed are small; therefore, the control loops are tuned to respond slowly. However, when a pump is started, the flow to the tank changes instantly. This will cause the tank to overflow. A solution to this problem would be to provide a set of fast-response tuning constants to replace the slow-acting ones whenever tank level increases rapidly.

SUCCESS STEP 2—DETAILED DESIGN

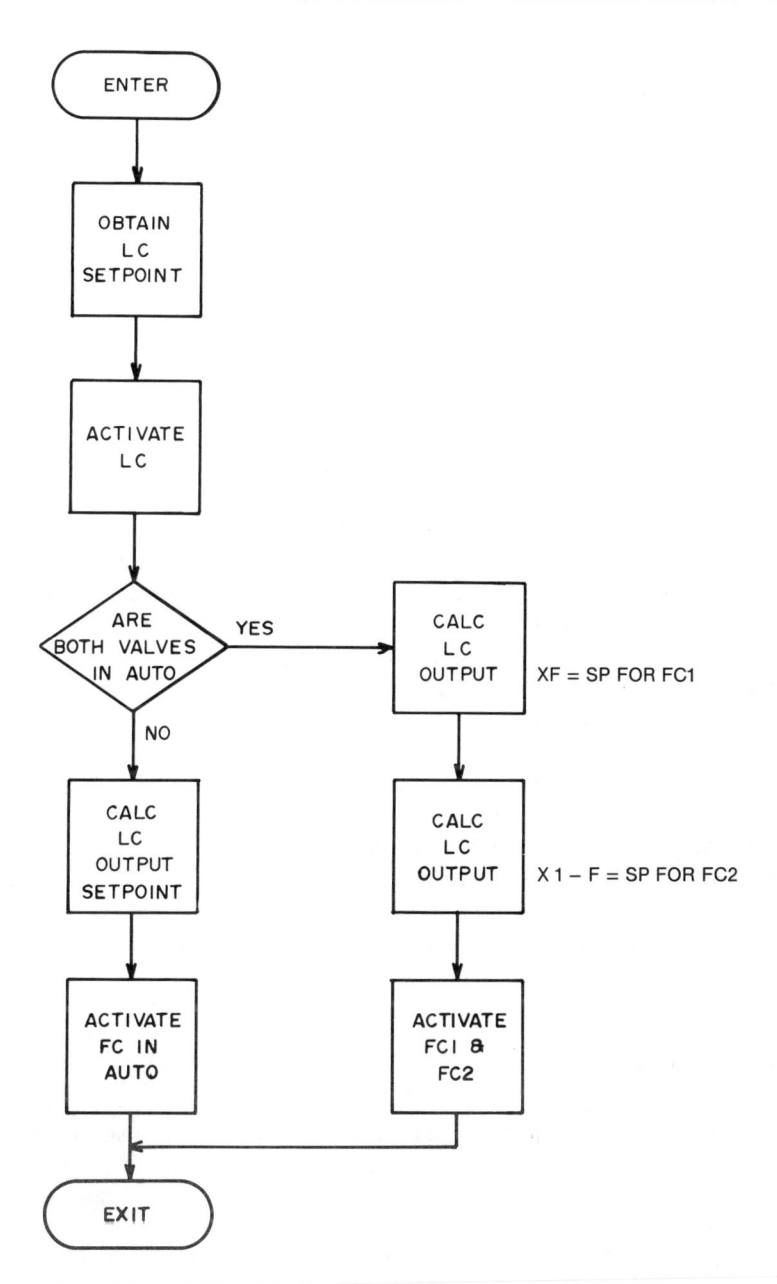

Figure 8. Simple flowchart.

Other software control modifications are possible and should be included if appropriate to the situation. Since robustness increases the software complexity and, hence, cost, it should be balanced by the expected benefits. Nuclear power plants are likely to have a high degree of robustness and so might many potentially hazardous or economically critical processes. However, for the most part, a high degree of protection is not necessary. Each process must be evaluated individually.

Analysis and System Design

Unit processes such as the tank and flow splitting example are just one segment of a plant or system. The units of a system will interact with each other to produce results that are determined largely by the control system. By careful planning of the system control strategies, the unit strategies should complement each other for optimum harmony. An example would be the advantage of a large-capacity tank to aid upstream or downstream processes. If the tank is small, then control strategies for units up- and downstream should compensate for the tank's limited buffer size. This example is very simple by comparison to most processes. In fact, a buffer is often inserted between two unit processes to smooth interactions and to make control simpler.

Every system should be analyzed to determine the compatibility of unit controls and unit processes. The analysis should employ the same methods used to develop a control loop. However, as the complexity of interactions increases, such analysis becomes subjective, eventually becoming a qualified guess. Analysis can be improved by rigorously applying systems theory or by simulation. Both are expensive to do and are usually used only on the most critical processes. The result of analysis should be to verify controllability of each unit and the plant as a whole.

Instrumentation

Selection and installation design of an instrument for each situation is an integral part of control strategy design. Low availability of an instrument will cause degraded control. Therefore, each instrument application should be reviewed to verify that the right type of instrument is selected, that appropriate optional features are chosen, that the installation is adequate for the instrument and that maintenance and calibration can be performed easily. An estimate of availability and reliability should be made. If they are low, mitigating efforts should be taken. Possible ways

SUCCESS STEP 2—DETAILED DESIGN

to raise performance include two instruments in parallel, an improved sample system, an improved installation for maintenance efficiency and a better instrument.

The same care given to instrument design should be given to the control devices: valves, gates and pumps. Size and control range (minimum position or speed to maximum position or speed), as well as availability and reliability, should be reviewed. Equally important is the interface between the computer system and the device. The interface should be designed to be fail-safe. For example, if the multiplexer fails, there should be no unwanted control actions. In most cases, no change in the control device is the desirable fail-safe state. A typical interface is shown in Figure 9.

Besides electrical compatibility, the interface must match the understanding of the information received with the information given. An instrument with a signal of 4–20 mA is calibrated over a set range, say 10 to 100 gpm for a flow transmitter. The multiplexer needs to know this to interpret the signal correctly. This and similar information for each signal (or point) in the system are kept in a document called the data base. The data base, like P&IDs, is a major coordination tool among the design team members. It is also a major coordination tool for the vendor during implementation and serves in this role throughout the life of the control system.

DESIGN OF DATA HANDLING

A digital control system can handle large amounts of data, which can be reported, stored and condensed in many ways. In design it is important for the user to consider what to do with these data. If provisions are not made during design, it is likely that desirable data will be missed. Furthermore, for large systems one could likely be overwhelmed by all the data. A rationale for data presentation must be designed.

Data can either be instantaneous (online) or historical. Presentation of instantaneous data and operator responses to it comprise a subject by itself. It is called operator/process interface (OPI) and is treated in the next section. Historical data are stored by the computer system, first on disks and sometimes transferred to tape for long-term storage. The manner in which data are stored, transferred and made available to the user is restricted by the storage size of disks and tapes. Storage procedures must be defined for each application to match the user's needs.

There are many forms and uses of historical data, but they usually are processed to generate reports and trends or plots. Processing can include statistical analyses, special calculations and special formats. Because of

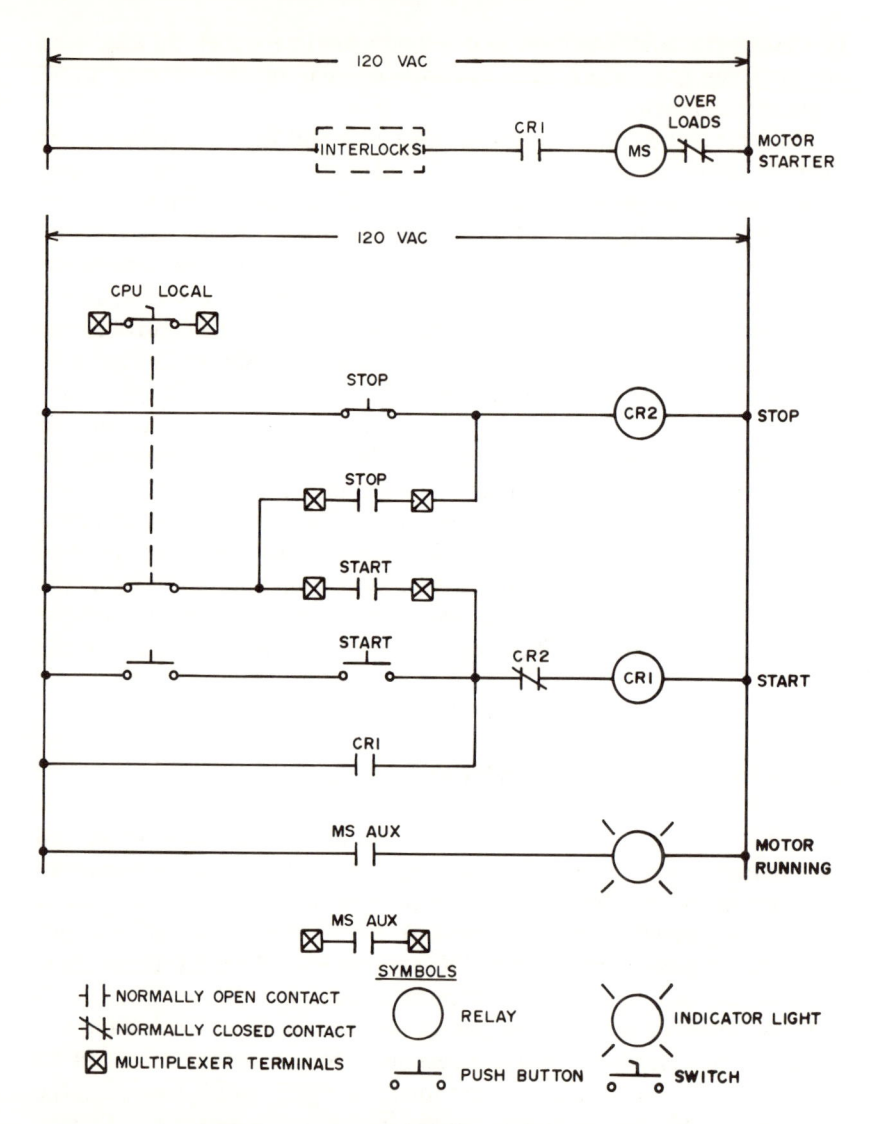

Figure 9. Wiring interface start/stop pump.

the almost infinite possibilities for report layout and organization, user input is vital. Without this input, little can be done by the designer except to define scope and specify general report compilers, thereby allowing report content and format to be defined easily by the user at a later date.

For most control systems, the data can be categorized as (1) process,

(2) maintenance, (3) laboratory, or (4) administrative. One possible data handling configuration is shown in Figure 10.

Process Data

Process data are the major interest in a control system. They are easily available from online instruments and their importance is readily appreciated by user and engineer alike. After all, the process is the *raison d'être* of the control system. Reports often generated for process summary include shift, day, month, quarter and annual operating reports. Logs of events and alarms also are printed.

Energy usage and plant productivity are very important aspects of process data. Calculations from online data can be made to report on performance in each area. Some important parameters easily reported include electrical energy demand and power factor, fuel consumption, energy balances, production rates per unit energy consumed, downtime and mass balances. Instrumentation to make these calculations possible is usually cost-effective because early detection and correction of problems will be possible.

Maintenance Data

Online instrumentation also can be used to keep track of information important to plant maintenance programs. Run times and frequency of use can be put into preventive maintenance schedules to enhance calendar time scheduling. For small systems, a weekly or monthly printout of such information may be all that is needed. For large plants, some form of automated maintenance scheduling is likely justified, probably with a separate computer. In any case, the data handling must be designed to meet the requirements, from report printing to intercomputer communication of maintenance information.

Laboratory Data

Lab data are used mostly for quality control checks and to monitor key process parameters. These data often are very useful to the process operator. Like process and maintenance data, lab data can be available from online instruments such as gas chromatographs, atomic adsorption units and others. However, a considerable amount of lab data must be

196 PROCESS CONTROL COMPUTER SYSTEMS

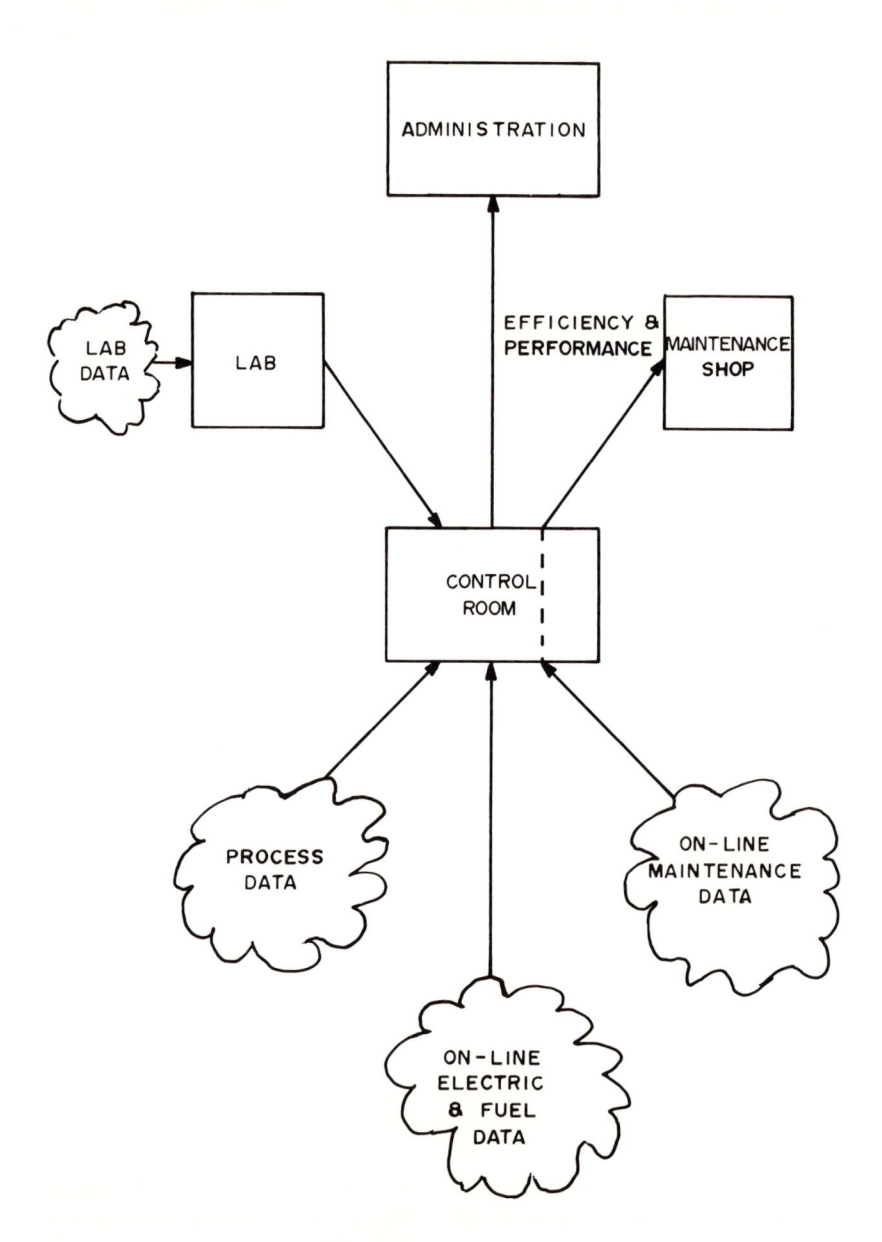

Figure 10. Data users.

entered manually into the computer, either directly by typing or from keypunched cards. If the data are entered in a raw form, then conversion to recognized parameters is needed. This can be handled by a separate laboratory computer. Again, in any case the data handling must be designed to meet the needs: from manual data entry to intercomputer communication.

Administrative Data

Administrative data handling is concerned with payroll, billings, accounting and special reports. Programs to process this information are performed offline, either on a backup computer or on a separate computer. Some special reports may have need of online data, which can be obtained from the process. Similar to maintenance and laboratory data, the data transfers must be designed. In addition, potential use of the backup computer needs to be included in the design effort because it could reduce system availability.

DESIGN OF OPERATOR/PROCESS INTERFACE

Any control system, regardless of size, is merely an aid to help the operator run plant equipment. The operator sets the direction for the control system. Although sophisticated control systems can overcome complex problems through contingency actions, the intelligence to run even a simple process resides in only one place: the operator. To be an effective tool, the control system must have a means to show process information and a means to accept operator instructions. This interface between the operator and the control system is ultimately between the operator and the process. For this reason, it is called the operator/process interface.

Design of the OPI is an important part of the system. It can greatly affect safety, operating costs and training. The design team must consider many things, among them how an operator takes in information, thinks and reacts. What makes an operator bored or alert, tired or fresh? The study of these things is called ergonomics, or human factors engineering.

In large plants, where the OPI is remote from the process, proper design of the OPI increases in importance. An operator in a control room depends entirely on the OPI for all feedback. Except for an audible alarm, all feedback is usually visual. Other senses — touch, hearing and

198   PROCESS CONTROL COMPUTER SYSTEMS

smell—that are used at the equipment are not used in the control room (an exception to this in some systems is the use of microphones to report sound). A good OPI should compensate for the sensory loss, which is done by presenting information logically in more than one way. A VDT can show tables or alphanumeric data, bar charts, trend charts and schematics or graphics.

VDT Displays

Graphics

The primary method of OPI is through graphics. Graphics are schematic representations of the process shown on VDT displays. From a graphic display the operator can obtain current information about the process and can make control changes. The "picture" of the process helps to organize data and speed operator action. The operator should be able to change setpoints, select automatic strategies and perform manual control, such as opening/closing valves, starting/stopping equipment and increasing/decreasing equipment speeds. An example of a graphic display is shown in Figure 11.

Graphic layouts should be prepared during design. Again, the biggest problem in layout during design is obtaining user input. Like the reports, graphics are a matter of personal preference. They require the input of the people who will use them.

Control Displays

Another method of OPI is by using alphanumeric control displays. These displays contain a list of parameters and their current values or status. The operator may select a parameter and change it as desired. Some parameters, such as controller tuning constants, may be removed from the operator's capability and reserved for a process engineer.

Control displays are not as easy for the operator to interface with as graphics. One way to improve control displays is to use bar charts to show relative values of parameters. Typically, design step layout of these displays is not done because standard off-the-shelf alphanumeric display capabilities are sufficient. An example of a control display is given in Figure 12.

A special type of control display is the interactive data entry (IDE) display. The VDT display prompts the operator on what to do next in response to entries and process conditions. An IDE might start with a list

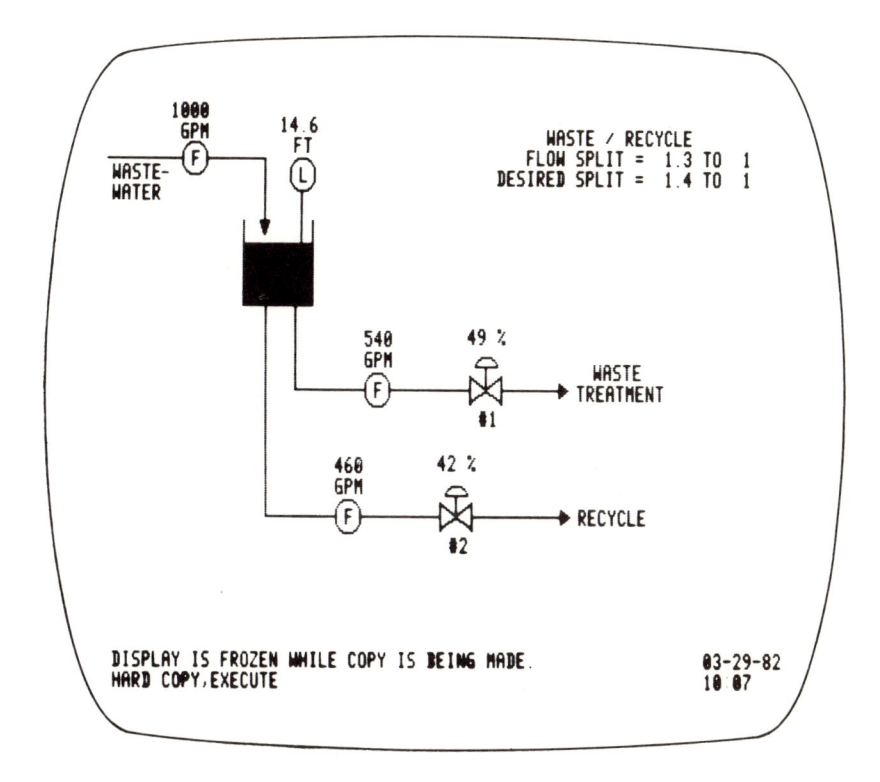

Figure 11. Graphic display.

of options and the values or status of important parameters. Based on the operator's entered action, the control system performs any control action required and displays the next step in the sequence. This interaction continues until the operation is completed or aborted. Defining and programming IDEs is difficult. IDEs therefore are reserved for complex operations where the benefit of IDE is needed. IDEs should be defined during design to give the vendor clear direction. An example is shown in Figure 13.

Summary Displays

Digital systems will contain many points, which may be in two or more states. Searching all points on control displays, graphics or other displays looking for points in a particular state could be very time-consuming. Therefore, a display that lists all points in a particular state can be defined that will allow rapid review by the operator.

PROCESS CONTROL COMPUTER SYSTEMS

```
WASTEWATER FLOW SPLIT
     TOTAL WASTE FLOW              1000 GPM
     SPLITTER TANK LEVEL            14.6 FT
     SPLITTER TANK SETPT            13.0 FT
     SPLITTER TANK CTLR             AUTO
     SPLITTER TANK CTLR OUTPUT      70%
     WASTE TR FLOW                  540 GPM
     WASTE TR FLOW SETPT            530 GPM
     WASTE TR CTLR                  AUTO
     WASTE TR CTLR OUTPUT           49%
     WASTE TO RECYCLE               1.3 TO 1
     WASTE TO RECYCLE SETPT         1.4 TO 1
     RECYCLE FLOW                   460 GPM
     RECYCLE CTLR                   MANUAL
     RECYCLE CTLR OUTPUT            42%
```

Figure 12. Control display.

Probably the most important summary display is the one for those points in alarm. The display will show all points with unacknowledged alarms, those that were acknowledged and other categories of alarms. Other summaries include computer/local status, equipment out-of-service, failed instruments and instruments off-scan. Capabilities of the OPI need to be defined for each summary.

Trends

A trend is a plot of parameters over time. Trends can be made on an analog instrument, such as a strip chart recorder, or on the VDT. Perhaps not unexpectedly, the principal use of the plots is for observing trends. Some vendors have adequate trend programs but the requirements for trending depend on the user and should definitely be part of the design.

Keyboards

The keyboard allows an operator to control the plant and get information. It must be designed. The control philosophy of the system should set the keyboard. For a plant system, the keyboard must be powerful to

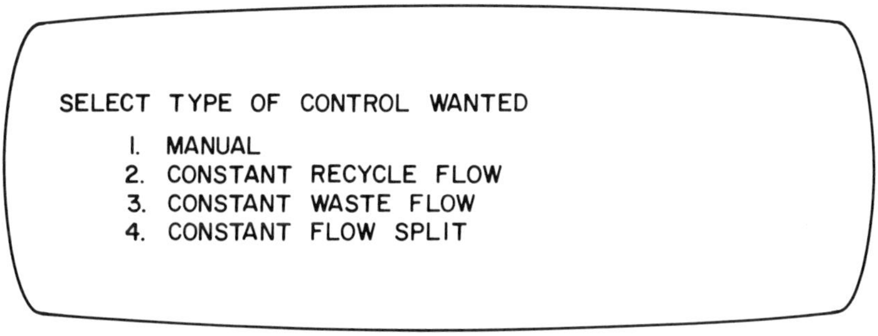

SUCCESS STEP 2—DETAILED DESIGN

```
SELECT TYPE OF CONTROL WANTED
    1. MANUAL
    2. CONSTANT RECYCLE FLOW
    3. CONSTANT WASTE FLOW
    4. CONSTANT FLOW SPLIT
```

Based on operator selection, the screen will respond with appropriate instructions.

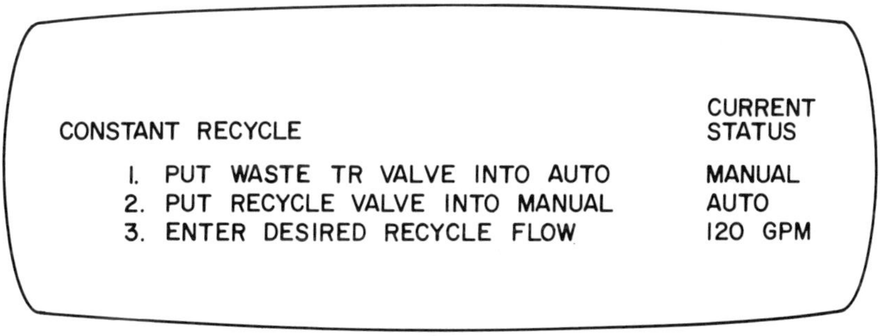

```
                                            CURRENT
CONSTANT RECYCLE                            STATUS
    1. PUT WASTE TR VALVE INTO AUTO         MANUAL
    2. PUT RECYCLE VALVE INTO MANUAL        AUTO
    3. ENTER DESIRED RECYCLE FLOW           120 GPM
```

Computer verifies proper operation; for example, if the operator put the recycle valve into manual without the other valve being in auto, the operator would be given an alarm indication and reason for alarm.

Figure 13. IDE display.

do many things. On the other hand, a keyboard for a system running one piece of equipment may only need to do a few things. In both cases, the keyboard should be simple and allow for fast operator action.

Of the keys on the board, the ones in most need of design are function keys. Function keys cause the system to do a specific task. A "DISPLAY" key, for example, will cause a chosen display to appear on the VDT.

Choosing the display is another step using other keys, cursors or light pens. In a large system this may not be fast enough for all cases. To speed up selection, other function keys may be added, such as "DISPLAY ALARMS" or "DISPLAY TREND."

Non-VDT Displays

Besides the VDT, there are other ways to help the operator observe and control the plant. These should be used as needed to make the right OPI. Again, "right" means what matches the user needs and resources. Some of the items that should be considered in the design are as follows:

- Graphic panels
- Recorders
- Printers
- Digital readouts
- Horn and annunciator
- Closed-circuit television
- Voice messages
- Video copiers

Of these, the two most important are printers and video copiers. Both give an operator a piece of paper that can be studied, saved and marked as needed. OPI printers produce logs of alarms and events. Other printers are used to make the less frequently needed reports.

Operator Environment

In addition to the displays and the keyboard to interact with them, the surroundings of the operator should be designed to reduce distractions and fatigue. Aesthetic appearance of the control room and equipment, work space, spatial arrangement, lighting, noise, heating and ventilation all contribute to the operator's performance. Of these, good lighting without VDT screen glare is the most often overlooked. Communication, power and fire safety systems need to be planned for the control and computer rooms. All these factors should be included in design of the control room.

SPECIFICATIONS

The product of design is a documented specification of the control system. This document's most important goal is clarity. If it clearly defines the system, then the task of implementation will be straightforward. If it is not clear, time and money will be lost "redesigning" ambiguous portions. This has been a very significant cost on some projects. Assuming the steps for successful design have been followed prior to writing the specification, the only step left is to make the specification clear.

If the system is to be provided by a vendor on a turnkey basis, the specification must include the technical portions describing the system, the contractual ("boilerplate") portions and detailed definition of schedules, submittals, testing, demonstration and system support. Vendor requirements in these areas allow persons managing the implementation of the system to do their job.

Just as in any construction activity, certain portions of the work must be completed before others can be started. The specification should set major milestones, set the order in which shop drawings are submitted for review, and set conditions under which factory and field tests can be conducted. When writing a specification for procurement from an outside firm, it is particularly important to tie the milestones, submittals and tests to monetary restrictions to give some power to the construction manager. Finally, the last payment should depend on successful demonstration of the working system. Again, in each requirement it is essential to be clear because the specification is a legal document.

The contents of the specification can be organized in several ways. One possibility is outlined in Table II. One part of this outline, "system performance," was referred to indirectly as a design option. To a certain degree all computer system specifications will be performance-type documents because each vendor has a different method of performing the same functions. Some of the parts are nonstandard, including configuration, special software for OPI, and data handling and algorithms. Other parts may be "off-the-shelf." The parts must work together to meet performance standards. To obtain the desired performance, it is essential to specify it.

Even if a specification is to be implemented in-house on hardware procured directly on a purchase order, the same level of detail is required from a technical point of view. The only difference is the "boilerplate" and some of the general topics, such as system support, submittals, etc. Regardless who does the work, the details must be thought out and written down.

SUMMARY

Design is the only sure way of getting the control system desired. Whether the system is small or large, its design must be given careful

Table II. Typical Specification Outline

	Project Definition and Objectives
100	General 　　Schedule 　　Submittals 　　Payments and tests
200	Hardware
300	Software 　　Including OPI and data handling
400	Algorithms 　　Control strategies
500	System Performance 　　How the software and hardware parts fit together
600	Operational Availability Demonstration 　　Demonstration of the completed system
700	Support 　　Control room environment 　　Final documentation 　　Spare parts and test equipment 　　Computer room furniture 　　Maintenance contracts

consideration and must be complete. Starting with user needs and designing down to devices and software programs will provide the framework to meet system goals. With this in mind, the following do's and don'ts are recommended:

Do's

- Do be organized.
- Do invite user input.
- Do assemble a well-balanced design team.
- Do prepare and follow a *Design Guide*.
- Do establish freeze dates.
- Do prepare P&IDs early.
- Do identify control problems early.

- Do simulate critical processes.
- Do layout graphics.
- Do define IDEs.
- Do define report capabilities.
- Do define data handling capabilities.
- Do define system performance.
- Do be clear in writing specifications.
- Do plan for all aspects of the control system.
- Do prepare complete specifications, even for small systems done in-house.
- Do provide robust system configuration and algorithms.

Don'ts

- Don't hire a vendor to do system design.
- Don't design parts before designing the system.
- Don't assume something will be provided if it is not specified.
- Don't take design shortcuts.
- Don't overlook control device sizing.
- Don't over-design algorithms.
- Don't design controls after process equipment design is frozen.

CHAPTER 9

SUCCESS STEP 3 – IMPLEMENTATION AND INSTALLATION

Timothy P. McConville

INTRODUCTION

This chapter defines the events, responsibilities and some of the problems to be encountered in the implementation and installation of a process control computer. Each system will be unique, yet the steps required to assure success are very similar. Application of the ideas in this chapter is necessary to protect the investment in a particular system and ensure one gets the expected results.

During Step 3, system concepts and design will be converted from ideas and drawings to the reality of a process control computer system. For the transition from plan to actual system to be successful, careful monitoring and review must persist throughout this period.

Successful Step 3 work will result in a control system that will meet expectations and function smoothly. Although no system is trouble-free initially, this system will be operating at full capability sooner and will require less modification if the proper effort is applied during the implementation and installation period.

If attention is not given to the Step 3 responsibilities outlined here the result will be a system that fits the supplier's interpretation of the needs, not the manager's. Design problems will become system problems. More field modifications will be required to enable the system to operate at full capability.

VENDOR-SUPPLIED VERSUS IN-HOUSE IMPLEMENTATION

The system may be implemented totally by a vendor or by the engineering and programming staff, utilizing purchased hardware components. How it is accomplished depends on the technical resources available. Unless one has a strong technical staff with experience in a process control system of the same magnitude being planned, it would be best to enlist outside expertise. Whether the system is implemented in-house or by a vendor, most of the review processes and steps outlined in this chapter still should be followed. In either case, the same end result will be achieved.

With the in-house system, the roles of owner and vendor are played under the same roof. The project staff must be structured so that both roles exist. The review and quality control functions still must be present. However, coordination and communication often become easier.

The remainder of this chapter is structured as a vendor-supplied system. If in-house staff has implementation responsibility, a similar approach is needed, although the steps need not be followed as rigorously.

STEP 3 — GETTING STARTED

The Schedule

Whatever the size of the system, a detailed schedule of implementation and installation is required. The schedule becomes the management tool to gauge progress, plan workloads and anticipate upcoming problems. The larger the system, the more important and complex scheduling will become.

The owner or his engineer should be the keeper of the schedule. With an in-house system, the department that will be the end user should be charged with scheduling. The system specification or implementation plan should provide a reasonable time for implementation and definition of major milestones. At the start of implementation, the vendor should be required to submit a detailed schedule of activities. The vendor's schedule then should be incorporated into the master schedule.

Each milestone should be tracked to completion. If significant deviations from the schedule are encountered, the entire schedule must be reviewed for impact. Changes in equipment or scope also may necessitate a schedule revision.

Problems can be anticipated by tracking progress versus the schedule.

Often, time problems can be anticipated before the milestones are missed. When multiple contracts or departments are involved, schedule problems must be coordinated between all parties.

Responsibilities

Most of the responsibilities for implementation and installation should be defined in the system specifications. However, some responsibilities must be defined after the vendor is selected and implementation is underway. Both the vendor and owner should, as much as practical, identify which staff member will be responsible for each portion of the project so that communication may be directed appropriately. Correspondence and review paths should be clearly defined at the beginning.

User Participation

Active participation of the user should be encouraged from initial conception to system startup. Implementation is no exception. Important decisions and compromises may be made in this phase that will shape the final system. Areas of importance are the parts of the system that affect daily operations such as graphic displays, naming conventions, report formats and the like. This may be the last chance to shape the details of the operator interface so that the result suits the need.

Some areas, such as reports and graphics, require extra attention by the user. These are the tools the operator will be utilizing the most and that will heavily affect operator acceptance of the system. All of the operator interface definition should have active user participation early on and throughout the implementation phase. Even though graphics and reports were defined during the design phase, customization during implementation will be necessary due to the specific vendor's graphics package or changes in equipment or operating philosophy.

Correspondence

If correspondence paths have not been defined in the specification, they should be agreed on shortly after bid. Every situation will be somewhat different, depending on how many contractors, subcontractors and engineers are involved. The bigger the system, the thicker the paper maze.

The importance of a good correspondence tracking system cannot be overestimated. On larger projects, with a greater volume of correspondence likely, a computerized tracking system may be warranted. Use of serial numbers on letters and good letter subject descriptions helps considerably in tracking letters. Both will facilitate adaptation to a computerized tracking system.

Another often ignored area is a record of telephone conversations. Much information is exchanged and many decisions are made via telephone. If good records are not made of phone conversations, an important part of the job history will fade into the memories of the callers. Failure to document phone records also may lead to misinterpretation and misunderstandings. Agreements made by phone should be documented by letter.

When writing letters pertaining to the contract, keep this in mind. If claims are made during the course of the job, all correspondence is part of the legal record. Write letters accordingly. At the beginning of the job, when everything is going smoothly, it is hard to remember this. However, the job does not always end as smoothly as it begins.

Transmittal Procedures

Before the first submittal is sent, agreement should be reached about its content and format. Everybody involved should know what is to be contained in the submittals and what the review process will entail. Some of the areas of potential problems are as follows:

- Level of detail
- Format
- Review and turnaround times
- Meaning of review comments
- "Revise" versus "approved as noted" comments
- How many copies and where they are to be sent
- Processing changes

If items such as these are discussed and agreed on before the submittal process starts, the process will be smoother.

SHOP DRAWING REVIEW

Review of shop drawings and contractor submittals is an important and time-consuming part of the job. Computer system contracts usually have software as major components of the job. Software is not tangible and readily observable. Inspection of the contractor's work takes on a dimension somewhat different than would inspection of a building contract. Even when implementing one's own system, design documents from the programming staff are absolutely necessary and the review just as time-consuming.

The review phase of the contract allows the engineer and owner an opportunity to check on progress as the system is built. During these checks, contractor and design errors can be corrected before they are built in. Contractor submittals should cover all components of the systems—both hardware and software. Functional capability of all components should be described and checked.

Preparing for Review

Hopefully, the specifications have defined the items for which the contractor must submit. If not, a list should be agreed on as soon as possible. The review process also should be agreed on by all parties. As soon as practical, the contractor should be required to provide a schedule of shop drawing submittals and expected submittal dates.

Before the review process begins, one must be prepared for the review. One must decide ahead of time who on staff will be responsible for each type of submittal (e.g., electrical, hardware, software, algorithms). An internal quality control procedure should be set up to ensure that the submittals are reviewed properly and there is coordination among the different staff members. Timely review, whether of a contractor submittal or in-house design documents, is critical. Lost time often affects the critical path.

Operator/Process Interface Definition

The operator/process interface (OPI) portion of the system will affect all algorithms and some of the software. Because of this and its importance to the system, the interface should be agreed on early in the review process. It is highly probable that the system supplier has a somewhat different standard operator interface than what was specified, although functionally equivalent. Usually a compromise can be reached. When compromising, remember that the operator interface is the tool that the operator must use day in and day out. If these tools are cumbersome or difficult, system acceptance will be poor.

In coming to agreement over the operator interface, conventions for displays and access must be defined. For example, with color graphics one must agree on a symbol set, i.e., whether a running motor should be green, red, etc., whether alarms change colors, flash, make noise or all three. These examples are just a few of the many conventions that must be defined. Also, the keyboard layout, types of displays and access to displays must be agreed on. The important things to look for are ease of use, flexibility and full range of the desired functions.

Software Definition

Software review will be from a functional viewpoint. It would be highly impractical to consider reviewing the actual software coding. It will be the implementor's responsibility to ensure that his/her own detailed coding functions properly. However, software concepts and organization should be reviewed as the job progresses. Checks should be made so that important functions or programs are not overlooked until implementation time. The process control language (PCL) should be examined carefully to ensure that all the intended functions can be performed. Familiarity with the vendor's PCL will be extremely useful during algorithm review.

Loop Drawing Review

As the control system panels and field interfaces are defined, the wiring or loop drawings must be reviewed. This may well be the last time to catch interface problems before construction. Loop drawings should be submitted in detail. All panel and field devices should be shown, as well as all wiring and relay logic. Although the prospect of a detailed review of each loop is impractical, review should not be taken too lightly. As a minimum, review should include cross checks to the following.

- **Input/Output (I/O).** All data base I/O points should be checked for missing or extra points.
- **Panel devices.** Panel devices should be checked against the process and instrumentation diagram (P&ID) and panel drawings.
- **Field device.** Devices should be checked against P&ID and data base.
- **Loop description.** Descriptions should be checked for agreement with data base and panel nameplates.
- **Instruments.** Calibrated range should agree with data base and instrument schedule. Any special calibrated dials or indicators also should be in agreement.

- **Wiring**. Typical device wiring diagrams should be checked for conformance to the design drawing and for compatibility with supplied field equipment.

Changes of field equipment or suppliers are of particular importance. Equipment changes may have necessitated wiring or I/O changes for proper interface. Review should be more detailed in the areas in which equipment changes have occurred since the original design.

Wiring and relay ladder logic review can slow down the loop drawing review. The load in this area may be lightened by only reviewing typical control circuits. Usually, a large number of loops can be reduced to a small list of similar loops. If only a few of the similar loops are checked, the review can be shortened considerably without much sacrifice of quality.

Often, when faced with an overwhelming pile of wiring diagrams, the reviewer is tempted to reach for the approval stamp as a quick means of clearing his desk. However, the reviewer should resist the temptation and check before using the approval stamp.

Algorithm Review

Control algorithms or strategies are the heart of the system. Most of the real-time decisions regarding process control will occur in the algorithm logic. Final definition of displays also depends on information generated by the control algorithms. The amount of review work required will depend very much on the complexity of system algorithms.

Before control algorithm review begins, the PCL should be approved and electrical interface (loop drawings) fully detailed for the controlled field devices. If the electrical interface is defined, the impact of field changes to the control algorithms will be easier to assess.

If not defined closely in the specifications, format and content of the algorithm submittals must have been agreed on previously. They should contain at least the following: a clear functional description of the algorithm, PCL implementation of the algorithm, all calculations, all operator displays, interface to other programs and the field inputs and outputs.

Although the algorithms should be reviewed for functional conformance with the specification, understanding of the PCL implementation is important. Such level of review requires the reviewer to be familiar with the system's PCL. It is important that the reviewer acquire the needed familiarity to understand the PCL implementation.

When reviewing the algorithm submittal, the reviewer should check

214 PROCESS CONTROL COMPUTER SYSTEMS

for conformity to the specification, uniformity to other submittals and conventions and proper control of field equipment. A checklist such as shown in Figure 1 will aid in the review process.

PROJECT _____

SUBMITTAL _____

OBJECTIVES:
- Ensure consistency of review
- Ensure timely review
- Ensure compliance with the specifications
- Ensure compatibility with the hardware

PROCEDURE:
1. Complete the checklist. Each item must be checked off and comments recorded.
2. Prepare test procedures for final testing (see end of checklist).

CHECKLIST

A. CONTROL GRAPHIC

____ Is the piping correct? (See P&IDs) _____

Are all of the following shown correctly?

____ Upstream/Downstream Equipment _____

____ Devices (pumps, valves, etc.) _____

____ Sensors (alarm, CM, etc.) _____

____ Calculations _____

____ Targets _____

____ Are all related displays accessible? _____

____ Overviews required? _____

B. I/O POINTS

____ All related points referenced? _____

____ All referenced points exist? _____

____ Are units correct? _____

C. STRATEGY

____ Is original spec still valid? _____

____ Does strategy match field hardware? _____

____ Are all wait times reasonable? _____

____ Are all sequences proper? _____

____ Are all specified control modes included? _____

____ Is strategy activation/deactivation clean? _____

____ Does flowchart match the description? _____

D. CONTINGENCIES

Review the entire strategy; have the following contingencies been addressed?

____ Sensor failures _____

SUCCESS STEP 3—IMPLEMENTATION

    ____ Control device failures _____
    ____ Devices unavailable _____
    ____ Mux failures _____
    ____ Alarms _____
    ____ Power failure _____
    ____ Failover _____
    ____ Alternate strategies in case of failure _____

E. TUNING

Are all of the following accessible for easy tuning?
    ____ Constants_____
    ____ Rates/Ratios _____
    ____ Time delays _____
    ____ Gains _____

F. CALCULATIONS

    ____ Are all calculations addressed and correct? _____
    ____ Are all available without the strategy? _____
    ____ Are all necessary data available? _____

G. MESSAGES

    ____ Are all necessary messages addressed? _____
    ____ Is the text of each clear? _____
    ____ Are common messages logged only once? _____
    ____ Are vital messages logged periodically? _____

H. USABILITY

A significant problem often encountered in the field during startup is related to the ability of the operator to change from one control mode to another and to execute the strategy with something less than a full complement of plant equipment (e.g., only one thickener out of four available).

    ____ Are mode changes spelled out? _____
    ____ Are they easy to do? _____
    ____ Can the strategy be run with less than full equipment? _____

Figure 1. Control algorithm review.

FACTORY TESTING

When all system submittals have been approved and the system is ready for installation, the owner must test the system. This test provides an opportunity for the owner to test system performance and identify the problems before delivery. The preshipping test is in the best interest of all parties and should not be underrated or bypassed.

By having the contractor demonstrate the system before delivery, the owner will know exactly where the contractor stands relative to the schedule. It also precludes delivery of a half-developed system. The contractor will benefit by having the owner find problems while the system is still on the factory floor. This allows the contractor to fix problems while it is convenient. Costly onsite time and expenses can be saved. Identifying and correcting system problems prior to installation also will provide for a less troublesome field startup.

Basic requirements for testing the system should be described by the specification. Details of the tests are the supplier's responsibility. The supplier should be required to furnish the test schedule and procedure. Obviously, the procedure must meet approval of the engineer and owner before testing can proceed. The factory test can become quite involved. A week-long test probably would be sufficient for a moderate size system of 2000 I/O points; however, one should not expect to be able to test all details in that short a time.

Test Plan

The test plan should be as detailed as possible and listed on a form that allows items to be checked off when tested. Insist on sufficient spacing on the forms to take notes as the test proceeds.

Time will not be available to test each point. However, each type of system response and system function should be checked. Broad descriptions, such as "demonstrate lime feed graphics," should not be allowed. A lack of detail in the test means details will be missed. A very specific checklist makes for a more orderly and comprehensive test activity.

System testing should be as comprehensive as possible and should cover the following items:

- Data base generation and maintenance
- I/O scaling and conversion
- Operator displays and programs
- Reports
- Process control language
- System response at various task loadings
- Utilities
- Special programs
- System disk and memory utilization

Everything should be demonstrated to the manager's satisfaction. All failures, questionable responses and deviations should be noted.

Test Results

On completion of the factory test, one should have a clearer understanding of the system, a list of items for retesting and a list of items untested. The manager may even have a list of suspicious items that gave unusual responses or that were not fully understood. These will have to be checked as soon as possible.

The punchlist of items that failed the test can be separated into two categories: (1) those that must be retested prior to shipment, and (2) those to be retested before field operation. Separation should be based on the severity of the problem and its effect on the rest of the system. Key problems should be rectified before the vendor is allowed to ship the system. An important item such as an unfinished trending package should be retested prior to shipment, whereas a mixup between two alarm points is minor and can be left to be fixed in the field.

After all the lists have been compiled, plans should be made to retest failed items. The contractor and owner/engineer both should be allowed a reasonable time period in which to reschedule the test if the result dictates it. Plans also should be made for items that were not tested. If the list is not extremely long and full of important items, retest may be delayed until after installation.

INSTALLATION AND INSPECTION

Successful field checkout will depend on how well it is planned for and how well it is organized. Prior to start of field inspection, the plan for checkout and recordkeeping systems must be in place. One should know all that needs to be done before starting. Inspection will cover the following areas:

- Interface
- Instruments
- Panels
- Computer I/O
- System operation

Interface

Inspection of the termination of all field I/O points to the panel or multiplexer is vital. Equipment submittals should be checked for agreement of wiring. If all wiring is verified to the computer termination strips, checkout of the computer will be much smoother. Many times, wiring from the field to the point termination will be the responsibility of one contractor, while wiring of the computer system I/O may be another. Separate inspection of wiring is more important in these instances to identify responsibility, should problems arise.

Instruments

All instruments should be inspected for proper installation and mounting. One should be sure to check proper grounding as this is often a problem and make certain that owner and maintenance manuals are provided for all instruments supplied. Instruments should be calibrated and tested prior to process startup, and one should check to make sure that the supplied instrument meets the requirements for the ranges required by the process.

Panels

Panels should be inspected and tested. Panel submittals can be compared to ensure that all required panel components are included and properly located on the panel. Nameplates should be checked for proper title, spelling and locations. Each panel indicator and switch also should be checked. Lights should be checked to ensure proper wiring. Switches should be checked to ensure proper function, i.e., starting correct motor or opening a valve instead of closing. Panel indicators should be checked for proper range and functioning, e.g., does the panel gauge read the same as local gauges across the entire range. As-built documentation should be verified for each panel.

Input/Output

Each of the computer I/O points should be checked to ensure proper termination and software linking. Inputs should be checked against the field device and panel for uniformity of response. Analog inputs should be checked for scaling, filtering (if necessary) and alarm values. Data base and displays both should be checked for agreement. If messages are associated with change of value, logging should be verified.

Outputs should be checked for calibration as well as proper linking.

Momentary outputs should be checked to ensure correct pulse times and adequate resolution between pulse durations. Inspection at one site revealed that the vendor was capable of outputting up to 10-second pulses; however, only integer outputs were allowed. This was totally unacceptable and required a rewrite of the output software.

Testing should be done from the operator interface completely through to or from the end element. A coordinated team is required, with hand communications and all necessary drawings. One should check each point carefully. One should not accept simulations, samplings or tests of parts of the wiring for a single point at different times. There is no substitute for thoroughness at this point.

System Operation

During the field inspection phase, system function and problems should be observed. Although the system functions were tested during the factory test, shipping and installation may have had their effects. The complex and interrelated portions of the system are such that a minor change to correct one problem actually may cause some seemingly unrelated errors to surface.

As the field inspection phase progresses, records should be kept of all activities. Each of the checkout areas discussed previously should have checksheets to record checkout dates and problems. Punchlists of all problems encountered should be kept and updated frequently. This period is also a good time to test items that were left on the punchlist following factory testing. Any items that appeared suspicious during the factory test now can be checked out more thoroughly.

One should be thorough during this phase and be satisfied that everything works properly. One should not take the vendor's word that it works. It should be tested personally.

TRAINING

Proper operator training is always of utmost importance. The installation of a computer control system does not diminish the importance of good training. It actually magnifies the need. The training program should encompass all the staff. From operator to managers, everyone must become involved. The new computer system is only a powerful tool if it is applied and understood properly. Without a good training program for all associated with the system, success will come slowly.

Operator

Training for the system's operators should be done in two parts. One part should include the rudiments of the process to be controlled; the other will pertain to the mechanics of using the control system.

Process Training

Whether the system is part of a new plant or a retrofit of an existing facility, process training is important. Often, the main control room of a computer system is removed from the process for environmental reasons. The operator is then dependent on the computer's inputs rather than human senses to detect process changes. Since they are not able to physically sense the process, operators need a clear understanding of the process to fully realize the impact of computer-directed changes.

A training course is needed to familiarize the operator with the basic knowledge of the process interactions, process dynamics and at least a basic understanding of process control. This course also provides an ideal time to discuss plant procedures and new job responsibilities.

Ideally, the process training should precede the vendor training and include an overview of the system and what the controls will do. If the operators know what the computer is supposed to do, they will be able to understand the vendor presentation more readily.

Vendor Training

An extensive training course by the system supplier should be required by the system specification. This course should detail each step required to operate the system. The course should be as much "hands-on" as possible to facilitate system familiarity. Each keystroke required for operation and use of the process interface should be explained. The vendor course should address system concepts and hardware, as well as keyboard familiarity. Controls for each process area should be explained for both normal and contingency operations.

The vendor training often is the operator's first exposure to computer control. Care should be taken to provide sufficient time so that the presentation is slow, provides hands-on time and addresses the operator's concerns. The vendor must provide a clear, usable operator's guide as a reference manual for the presentation.

Software Training

Software training and documentation must be provided for the programmers, engineers or technicians who ultimately will be responsible for the system's operation and maintenance (O&M). This course should be designed to acquaint the student with all aspects of the system. Students should learn system operations and the steps required to operate all of the hardware. The vendor should provide hands-on experience with all application programs. System diagnostic also should be covered. Material covered throughout the course should enable the student to make modifications to the system and troubleshoot most system problems.

System Management

Training for the new system must not be limited to operating personnel. Management also should become acquainted with its new tool and understand its capabilities and limitations. Management also must become aware of the commitment of time, staffing and money needed to make the system a success.

This vendor presentation probably will require one day. An overview of the system and a demonstration of the operator stations would be expected. Of prime concern is management of the system and the data that will be accumulated. Managers should become acquainted with all the system logs and printouts. In addition, recordkeeping and data storage practices need to be discussed.

In addition to the vendor presentation, management seminars are needed to provide familiarity with computerized process control. In many plants, engineers represent the process manager level. In this situation, training may be required to provide the necessary knowledge of the details of computerized process control. Since the engineer is more likely to become intimately involved with the computer, a higher level of understanding is desirable.

If only a few engineers or managers are involved, sending them to an outside course may be the most cost-effective training method. However, if a large group requires training, bringing in instructors who can tailor the presentation to one's needs is most desirable.

Maintenance

Maintenance of the computer system hardware may present a completely new requirement or may fit well with an existing maintenance group. This will be very dependent on the level of hardware with which people are familiar. Although one may purchase a maintenance contract with the vendor, a good deal of maintenance will be required by plant personnel.

A vendor maintenance course should be required for system maintenance. Course length and level of detail must be tailored to fit one's needs. Following the training, maintenance personnel should be capable of using proper testing equipment and tools, running diagnostic programs, isolating trouble areas, performing routine maintenance of the computer and the peripheral devices and be as familiar with the software as required for hardware fault checking.

If the maintenance persons are new to digital control and require substantial training, long-term sessions would be desirable. If the training is spread out over a month instead of a week, the sessions will be far more valuable.

FIELD TESTING

Following installation and field inspection, a formal field test should be used to evaluate the readiness of the system. This test usually will be a repeat of the factory test. However, actual field equipment is used.

Certain minimum requirements must be met before the field test can be scheduled. All control loops must have been tested and all areas of plant control must be functional. Exceptions to this rule can be made if some plant areas are not functional as a result of problems that are not the responsibility of the system vendor. An example is a solids-thickening process in which the solids feed pumps are not functional due to use of an inappropriate pump for solids handling. Obviously, this plant area cannot be tested fully until the pumps are replaced. The computer system vendor should not be delayed in this case.

All outstanding items on the factory test and installation punchlists should be tested and removed from the list before field testing can be started. In addition, the system should be operating reliably without any recurring problems, which cause the system to lose control functions. The system should be required to operate for a period of 72 hours without serious system problems before allowing the system test to be scheduled.

The field test will be very similar to the factory test; however, instead of simulating control functions, actual control can be tested. Test definition and scheduling will require cooperation and coordination between the plant operations staff and the system vendor. Process limitations and safety must be taken into account when defining the test procedure.

During factory testing, a considerable amount of time was spent checking details of the graphic and simulation process responses. Field inspection should have verified those functions so that they need not be rechecked during the field test. More attention should be given to process control functioning and system performance. Items not examined previously, such as process interactions and algorithm functions, should be tested in depth. System response time and loading of the central processing unit (CPU) also should be examined in detail. With the entire system operating, video display terminal (VDT) response time may have diminished from that observed at the factory test.

Documentation

By field test time, most of the final system documentation should be complete. All documentation should be reviewed for completeness and ease of use. Operator manuals should be clear and geared for the operator. Step-by-step instructions should be given for all normally used functions. System documentation should be laid out logically and should include a good index. Enough documentation must be provided so that the user can solve problems quickly, without numerous calls to the vendor or management.

Documentation should reflect the as-built conditions. If panel and loop drawings have not been corrected by field test time, efforts should be well underway. Inadequate or incomplete documentation should be cause for delay of the field test.

Results

Following the field test, punchlists should be generated for all problem areas. Similar to the factory test, items should be sorted by importance. Some must be handled before the operational availability demonstration (OAD), which will be discussed shortly. Others can be handled during OAD. The list should include all tested items, documentation review and items existing on punchlists before the test. Most minor items can be handled during the OAD period. Any serious problems, such as faulty control programming or system instability, should be repaired prior to start of OAD.

Since one purpose of the field test is to check readiness for long-term owner use, one should not allow the system to go into the OAD period with major problems. The system supplier obviously would like to start OAD as soon as possible. If the vendor is usually slow to make corrections, the start of OAD can be used as leverage to get a faster response.

OPERATIONAL AVAILABILITY DEMONSTRATION

By the time the system passes the field tests, most of the problems should have been found and resolved. However, although everything appears to be working, one should not give final acceptance based on a single, one-time test of system functions. Long-term reliability must be demonstrated through an operational availability demonstration (OAD) period, which also serves as a good transition period from vendor to owner use.

Computer control systems are very complex, including hundreds and even thousands of individual, but related, programs. It is not possible to completely test all combinations of events. The only way to ensure that all "gremlins" are found is to use the system for a prolonged time. Sooner or later, all problems will surface. An OAD of six months is the minimum time required.

Requirements of the OAD should be clearly stated in the specification. Usually, reliability is defined as a percentage availability (e.g., 98%). Availability is calculated as

$$A = \frac{MTBF}{MTBF + MTTR} \times 100$$

where
A = availability
$MTBF$ = mean time between failures
$MTTR$ = mean time to repair

Defining Conditions

During the OAD period, definition of responsibilities is not always clear. Technically, the system has not been accepted and is therefore the vendor's responsibility, yet the owner is operating the system and providing most of the maintenance. This situation sometimes makes the vendor feel uneasy about the liability during this transition. The clearer the definition of roles before OAD, the better. The best way to handle this problem is to have the responsibilities defined as part of the bid documents.

Failure Reporting

Once definition of failure is resolved, a reporting system must be implemented. The best way to handle failure reporting is to keep a good

SUCCESS STEP 3—IMPLEMENTATION

log of all the system problems and corrections. The person filling out the log need not make the decision as to what constitutes a failure. Failure reporting forms later can be analyzed for system failures. Figure 2 shows a sample failure report form.

Punchlists

Current punchlists should be maintained throughout the OAD period to provide a means for both sides to monitor outstanding items. Progress also can be gauged by the evolution of the list. The OAD punchlist is merely a continuation of those generated during factory and field testing. If the list has been well maintained throughout the OAD period, definition of responsibility will be easier. This will become quite important at the end of OAD. If final acceptance and payment are held pending resolution of punchlist items, they will be fixed a little faster.

SUMMARY

Implementation can be reduced to a single descriptive phrase: unrelenting, painstaking attention to detail. Everything in the control system must be checked. Because of its complexity and interrelated components, everything must be rechecked following any change. There can be no substitute for diligence during the implementation phase.

In summary, the following do's and don'ts are recommended:

Do's

- Do insist on the same diligence, regardless of who does the implementation—vendor or in-house staff.
- Do insist on a detailed schedule early in the project.
- Do insist on frequent schedule updates.
- Do establish written procedures for controlling communications, changes and paperwork.
- Do insist on a detailed review of all design details, either via vendor submittals or in-house documents.
- Do check carefully for interface discrepancies.
- Do conduct formal factory and field tests, with detailed procedures and checkoff.
- Do perform complete and exhaustive I/O checks at each contract interface.

226   PROCESS CONTROL COMPUTER SYSTEMS

NO.

Associated CFR No.

1. Failure reporting:
    By: _____ To: _____
    On: _____ At: _____

   Contractor Notified
   on: _____ @ _____

2. Description of problem given: _____
   _____
   _____
   _____

3. Contractor's preliminary diagnosis of reported problem: _____
   _____
   _____
   _____
   _____

4. Contractor's description of actual cause discovered (be specific): _____
   _____
   _____
   _____

5. Contractor's corrective action (reference associated CRS, if any): _____
   _____
   _____
   _____
   _____
   _____

6. Repair started on: _____ @ _____ _____ _____
                                              Contractor   City
   (or four hours past the time of contractor notification; whichever is shorter)

7. Acceptance Check completed on: _____ @ _____ _____ _____
                                                   Contractor   City

8. Contractor's total minutes to correct: _____

9. Reviewed by city: _____ 10. Reviewed by field office: _____

11. Reviewed by engineer: _____ 12. Reviewed by contractor: _____

13. Agreed this was / was not a failure affecting operational availability:

    _____ _____ _____ _____   DATE: _____
     CITY   FIELD   ENGINEER   CONTRACTOR
                 OFFICE

WHITE: CITY / CANARY: FIELD OFFICE / PINK: ENGINEER / GOLDENROD: CONTRACTOR

Figure 2.   System failure report.

- Do keep meticulous records of failures, problems, tests and retests.
- Do require an OAD and then enforce the rules.
- Do insist on complete documentation and training.

Don'ts

- Don't assume anything, even if one's own staff is doing the work. Insist on written definition of their implementation.
- Don't allow the user to avoid involvement. Their input is vital.
- Don't "rubber-stamp" submittals or design documents.
- Don't overlook the details of the operator interface.
- Don't accept anyone's word that something works—check it.
- Don't test I/O on a sampling or "typical" basis.
- Don't assume that a "final test," conducted at one time, will catch all of the problems.
- Don't allow problems to be resolved verbally. Insist on rules, procedures and written documentation.

CHAPTER 10

SUCCESS STEP 4 – STAFFING

Grant L. Bennett, PE

INTRODUCTION

Success step 4 is to obtain a staff to operate and maintain the new control system. This chapter will consider who should lead the staff, what the staff does, how the staff should be organized, what qualifications are required and how to obtain the staff. Some considerations for abnormal situations and a list of practical recommendations complete the chapter.

THE SYSTEM MANAGER

Who is the System Manager

The system manager manages the people responsible for the operation and maintenance (O&M) of the control system. The number of people will vary with the size and complexity of the system, but a system manager will be required for any size system. The system manager's duties include management of the day-to-day activities, as well as setting up and carrying out overall system policies and procedures.

Why a System Manager is Needed

A control system is a tool to be used to help operate a plant. It is a tool that, if used correctly, will result in a well-run, efficient and cost-effective plant. The system manager ensures the correct use of this powerful tool and translates management and regulatory agency demands into operational policies and procedures.

Functions of the System Manager

Planning

This is the most important of the system manager's functions and must be the strongest aspect of his/her personality, experience and general capabilities. This person must be able to translate goals (management demands) into reality, while balancing budget, personnel, equipment and supplies.

Training Coordination

Many different types of training are required to ensure that plant personnel are knowledgeable and experienced in operating and maintaining the system. They are available from many sources.

Manufacturers supply O&M manuals when a system is purchased. These manuals should be kept up-to-date and be available to all O&M personnel. If properly written, these manuals describe in great detail how to find problems, identify faults, and fix and test the equipment.

On-the-job training is provided by senior-level personnel to junior people within the organization. This type of training promotes communication between personnel, and provides experience for the trainee in system O&M, and for the trainers in management and speaking skills. The system manager must coordinate the availability of trainers and trainees, training materials, and classroom space.

Outside agencies and organizations can provide specialized training when the needs are identified and justified by the system manager. Here, the system manager must coordinate budgets, availability of personnel, location of the training, and any materials and resources required.

Many times, training requires temporary changes in plant operations. The system manager must plan and coordinate these changes to avoid detrimental plant output or waste of plant resources.

The system manager must be able to identify where, when and how much training is needed and should ensure that personnel are trained in more than one specific area to be able to compensate for vacation, sick leave or other conditions affecting personnel availability.

System Change Review

The system manager must review and approve any changes to the system, changes that can result from management requests, regulation changes, equipment failures, abnormal demands on plant capacity, personnel changes, age of equipment, training requests, and process control strategy changes and experiments. The system manager must have contingency plans available for both short- and long-term changes — plans that can respond to immediate changes and to future requirements.

The system manager must review all changes to determine their effects on plant output, personnel requirements, equipment capacities, and availability of resources and budget.

Coordinating Hardware and Software

Both hardware and software have extremely important tasks to perform and they must work together to accomplish the system goals. It is the system manager who is responsible for ensuring that hardware and software do not work at cross purposes.

The hardware must be maintained periodically to make sure that it will always be available when needed by the software. This includes preventive and corrective maintenance. A properly coordinated preventive maintenance program will balance the needs of the system hardware with the availability of manpower and spare parts. Software must be prevented from making excessive demands on the hardware, which could lead to degraded performance or excessive maintenance.

Hardware technicians and software programmers must be coordinated to ensure meaningful results from their work activities. Blaming one another for system failures must not be allowed. Procedures should be established by the system manager so that troubleshooting can be done independently by hardware and software maintenance personnel to ensure quick success at isolating and repairing system faults.

Expansion of either hardware or software must be coordinated by the system manager to ensure that one will not overload the other. For the same reason, major modifications to the hardware or reprogramming of the software must be coordinated.

System User Coordination

The control system in any plant, when used to its full extent, can be used by several groups within the plant. The most important of these is, of course, the control and monitoring of the plant process equipment. Other important users could include (1) maintenance scheduling, including run time records, work records, parts expense records and reporting of these records periodically to plant management; (2) laboratory analysis calculations and reporting records; (3) communications with other plant control systems for coordinating parallel or serial plant processing; and (4) development of new or modified plant process control strategies for optimizing or changing plant processes.

The control system often is equipped with dual computers in the central control room. This facilitates the system manager's job of coordinating the various users because the backup computer can be used in a background mode by the users not directly concerned with controlling the plant. However, the system manager still must establish schedules and procedures for allocation of time and resources between the system's many users.

Vendor Interface

The system manager must maintain a good relationship with the vendor who supplied the system. The vendor is a valuable source of information concerning the maintenance of the system hardware and software. Many times the vendor will provide updates to the operating system software as they are developed. These updates usually either enhance the operation of the system or provide new capabilities.

The system manager should be on the vendor's mailing list for new product announcements to keep abreast of what the vendor is doing. Additionally, the system manager should make periodic arrangements for the vendor or the vendor's representative to inspect the plant and review how the system has been functioning. The vendor should be kept aware of any developing problems so that the vendor can be used as a source of troubleshooting information.

The system manager's relationship with the system vendor should begin as soon as the vendor has been selected and work has begun on the system. By keeping aware of all the development steps, the system manager will be better able to take advantage of the resources of the vendor when planning and scheduling for the system installation, testing, training and startup activities.

Supplies and Spare Parts

The system manager is responsible for ensuring an adequate stock of consumable supplies and spare parts commensurate with storage space, system needs and budget constraints. Consumable supplies include such items as printer's paper, printer's ink or ribbons, disk-cleaning tools, spare magnetic recording tapes and disks, file space, writing materials, technician's tools and materials, programmer's tools and materials, notebooks, punchcards, air vent filters, connector pins, wire-wrap pins, wiring, connectors, solder, soldering tools, storage cabinets for magnetic recording media, secure fireproof storage (or offsite storage) for backup system tapes and disks, and storage for logs and reports.

A secure and fireproof storage area for backup copies of system tapes or disk files is extremely important for continuous operation of the plant after any incident that may destroy or damage system memory. The system manager should keep close control of key access to this area.

The system manager is responsible for maintaining an adequate supply of spare parts to maintain the control system hardware, including a repair facility and the proper tools. The vendor should furnish a list of recommended spare parts on delivery of the system documentation, but the system manager will have to modify that list as system needs dictate. Therefore, the system manager must maintain records of spare parts usage to determine future needs and budgetary requirements.

Maintenance Scheduling

The system manager is responsible for planning and scheduling preventive and corrective maintenance on the system hardware and software. The maintenance plans must contain contingencies for corrective maintenance in emergency situations. The O&M manuals provided by the system vendor will form the basis for the system manager's basic plans, but the plans must be flexible to meet the special needs of the system.

Budgeting

The system manager is responsible for requesting and maintaining the operating budget for the system and the personnel under his direct supervision. A good balance must be maintained between the budgets for personnel and system hardware because too many parts with no personnel to install them is as bad for the system as too many people with no parts to install.

The primary objective of the control system is to maintain product quality and plant efficiency. Therefore, the system manager's primary objective is to keep the system able to meet its objectives efficiently and economically.

STAFF ORGANIZATION

Functional Responsibilities

Operations

The operations group is responsible for operating the process equipment in the plant. These duties may include the following:

- Replenishment of batch day tanks
- Observation of raw material usage rates and product quality
- Observation of equipment operating parameters to determine efficiencies and need for maintenance
- Replacement of full product trucks, tanks or bins with empties
- Recording operating events, such as alarms, equipment parameter changes, process changes, equipment failures, product parameters and process efficiencies
- Monitoring raw material supplies and reordering when necessary
- Unloading and storing new raw material supplies
- Packing and transporting finished products
- Cleanup of process equipment and operating areas
- Monitoring of plant utilities and arranging for adequate supplies
- Storing and reporting plant/process operating records
- Identification and diagnosis of plant equipment operating faults and failures
- Scheduling plant equipment downtimes and operating cycles
- Identification of control system hardware and software failures

Control System Hardware Maintenance

The control system hardware maintenance group is responsible for performing preventive and corrective maintenance on the system equipment. These duties may include the following:

- Lubrication of moving parts
- Cleaning of mechanical machinery
- Cleaning or replacement of cooling air vent fans
- Running hardware diagnostic tests to discover reasons for improper operation
- Additions and deletions of plant instrumentation input and output points, including rewiring of termination blocks, changing backplane wiring and adding or deleting point cards

SUCCESS STEP 4—STAFFING

Software Maintenance

The software maintenance group is responsible for diagnosing and correcting problems with the control system software. These duties may include the following:

- Finding and fixing latent "bugs" in the control strategies, which many times appear only after long use
- Modification of operating software and application software based on changes and updates from the system vendor
- Additions and deletions of plant instrumentation input and output points, including modifying scanning and alarm tables, logs, video display terminal (VDT) graphics, reports and messages
- Changes to process control strategies to accommodate process optimization, changes to process equipment, new process recipes, changes in raw material quality, and changes to plant instrumentation and control devices

Process Control Optimization

The system manager and staff have a never-ending job: constantly striving to improve the operation of the plant and process to make it more efficient and cost-effective, and to improve the quality of the product. Imperfections in plant design and construction, process equipment manufacture and installation, or raw material quality must be compensated for by improvements in plant operations. Once the engineers, contractors and suppliers have completed their tasks, it is left to the plant staff to tune out the rough spots and find the latent defects and discrepancies.

Some of the many ways of improving plant operations include the following:

- Varying raw material ratios in the various processes, both in laboratory experiments and in the actual process equipment
- Varying dosage rates of chemicals
- Varying reaction and soaking times of processes
- Varying speed of process mixing equipment
- Varying process measurements, both location and type
- Changing process equipment
- Changing raw materials
- Varying equipment or plant loading
- Seasonal changes to react to changes in process raw materials or ambient conditions

Relationship to Plant Size

The staff requirements for operating and maintaining a control system will vary with the size and complexity of the system. The existing staff and organization of the plant will also exert some influence, although it would be best to minimize this influence in favor of the system requirements.

It stands to reason that a large control system will require a larger staff but the relationship is not necessarily linear. Each system size—small, medium and large—should be considered separately in light of the unique requirements for each.

Large Systems

Large control systems [greater than 4000 input/output (I/O) points] will allow specialization of staff duties. That is, each person will be able to devote more of his/her time to just a few duties and, therefore, probably will become more proficient at them. This can be advantageous in that less learning and analysis time will be required to solve the problems in a person's particular specialty. In addition, rotation of personnel will expose everyone to the various parts and features of the system. One of the best results of rotation is the different viewpoint each person brings to the solution of a problem.

The organization of the large system staff therefore must be flexible enough to allow for shifting people around. Personnel should be rotated on the basis of duties as well as the day, swing, night and relief shifts. It will be necessary to make suitable arrangements and agreements with the applicable labor organizations. This flexibility will also facilitate shifting of personnel to cover for vacation, sick leave and training requirements.

Medium-Sized Systems

Medium-sized systems (1000–4000 I/O points) probably are the most difficult to staff. The size and complexity of the system require a substantial staffing level, but the limited budget of the plant restricts the freedom required to obtain the required number of people. It is most important here that the system manager be assigned responsibility for the system and be allowed to obtain periodic help from other plant personnel resources. Some overlap between positions must occur to compensate for limited numbers of budgeted positions. Rotation becomes more important than with large systems because fewer numbers of people will have to learn more aspects of the system.

Some examples of situations in which job overlays can be especially helpful are as follows:

1. Hardware maintenance people, who are at least familiar with how the software works, often can diagnose software-related problems. They also can fix simple software problems provided they understand the interrelationships within the programs.
2. Software programmers also can diagnose hardware problems if the programmers are sufficiently familiar with the hardware. In many cases, hardware problems are fixed by simple board replacement, which makes the fix less difficult than the diagnosis.
3. Process operators should become familiar enough with both the hardware and software to assist in the diagnosis of system problems.
4. System hardware and software maintenance people should become familiar with the plant process so that help can be given in the diagnosis of problems.

It cannot be stressed enough that training will play the most important role in ensuring that all plant personnel will be able to contribute to the smooth operation of the plant. Specialization is important but can lead to jealousy and overprotective attitudes when allowed to become too important.

Small Systems

The small systems (less than 1000 points) place the greatest demands on the staff as it has to perform many different jobs with limited resources. A well-run small system can, however, produce the highest benefits because the system will be able to handle the greatest percentage of the work of running the plant. Again, the key is to have a system manager in charge of, and responsible for, the system.

The small system staff must be sufficiently dedicated and self-motivated to take on the extra tasks and put out the extra effort. Particular efforts must be made by the system manager to keep job satisfaction and rewards at a high level.

Relationship to Type of System

The type and size of the control system staff will vary according to the type of control system. Two types of systems will be discussed: (1) the programmable logic controller (PLC); and (2) the custom-designed distributed direct digital control (DDC) system. In addition, the installation of a system into new and existing plants will be discussed.

PLC Systems

PLC systems are characterized by fixed-sequence programs and limited-analog loop controls. The software efforts required to program and maintain the PLCs usually can be handled by process operations staff with relatively little formal software training.

PLC systems often are applied to single processes. While there may be several PLCs in service throughout the plant, hardware maintenance can be handled by an instrument technician who is experienced with electronic circuits.

Distributed Direct Digital Control (DDC) Systems

The distributed DDC system will require a substantial effort by the control system staff. The areas requiring the most critical attention will be the communications link and the so-called global software programs.

The communications link typically is the most susceptible to problems, and the staff must pay close attention to ensuring isolation from noise sources and physical damage.

The global software programs are those that coordinate and control the various remote processing units that are each responsible for a section or a process of the plant. Changes or troubles with these programs can have far-reaching effects on the plant operations.

The distributed DDC system will require staff to maintain the parts of the system in the central control area, as well as out in the field at the remotes. Close coordination and communications between these two groups must be maintained. It will be helpful to rotate staff around to the various remotes so that each staff member learns about more than one process area.

The software staff must include at least one fully trained computer programmer who is intimately familiar with all aspects of the system software. The system software, as well as the applications software, must be learned. The software staff should be knowledgeable about the plant processes and equipment.

The hardware staff must be fully trained on the system hardware. The hardware staff will be dedicated to control system hardware only and will not have much time or opportunity to participate in plant equipment maintenance beyond the instrumentation and control devices. General familiarity with plant equipment will help in team problem solving.

Time should be budgeted for the hardware and software staff to implement vendor-supplied enhancements and modifications to the control system. These changes have been thoroughly researched by the vendor, and any benefits in system performance should not be delayed.

Installation in a New Plant

Installation of a control system into a new plant is, in many ways, more challenging than retrofitting an existing plant. All the plant equipment and processes will be new and untested, along with the control system equipment and software.

The nucleus of the control system staff, both hardware and software, should be available as early in the plant design effort as possible. At least the system manager, or the future system manager, should be available to assist and review the many system interfaces. Interfaces requiring attention are with the plant equipment, the buildings housing the system equipment, the plant processes, the instrumentation and controls, the operating philosophy, the maintenance philosophy, the system environmental controls, and the form of the system documentation. The system manager should participate in, and be satisfied with, the engineering study that justifies the economic and technical aspects of the system.

Supervisory staff positions should be filled during the construction stages of the system so that supervisors can participate in the review of the vendor's work. They need to be available for orientation-type training at this stage in the project. These positions and the people who are hired to fill them are extremely important to the success of the project. The investment of time and money that allows these future supervisors to live and grow with the system will be returned many times over.

The remainder of the control system staff should be hired for the detailed vendor-supplied training and to be available during installation, testing and startup of the system. This will help ensure a smooth transition between the vendor and the staff.

One danger must be avoided. The control system staff must not begin second-guessing the design details in the specification requirements. The staff should be familiar with the situation but should be concerned only when events occur that will make the plant difficult to control. During construction, the control system staff should be in an advisory position to the construction and inspection forces. The staff should be available for consultation but should not be a thorn in the side of the construction people.

Installation into an Existing Plant

Installation of a computer control system into an existing plant should involve the control system staff early in the planning and design stages. The staff may not even be identified as the control system staff at this point; however, early attempts to target and begin orientation of selected key members of the staff will pay off. The system manager should be one of the first selected.

In a retrofit project, the field instrumentation and control devices will require serious consideration of whether to replace or retain. Instrument maintenance technicians with experience on the devices can be excellent sources of information, as well as potential control system technicians.

Programming help for the new control system may have to come from outside, unless data processing programmers are available who can make the transition to real-time. Consideration should be given to existing personnel who are willing to be trained and promoted.

A great deal of planning must be done for a retrofit project. Among the things to be considered are scheduling of construction and testing activities so as not to affect regular plant operations adversely. This is probably one of the greatest challenges because there are so many factors involved, such as weather, coordinating contractors, late deliveries and production requirements. The control system staff will be very important and can contribute a great deal to the scheduling effort. When the scheduling and planning are done, the staff will be expected to complete the testing and inspection of the system functions within the allotted time.

The contractor supplying and installing the control system will apply a lot of pressure on the staff at this time because he has completed the major portion of his work and is anxious to be paid and leave the job. The staff must be prepared at this point to take over responsibility for the system, including operation and maintenance. The vendor likely will provide some additional consulting on especially difficult problems. For the most part, however, the staff will be doing most of the work.

One advantage of installing a control system in an existing plant is that the processes are probably up and working prior to installation. Therefore, the various parts of the process can be turned over for control and monitoring one at a time. This will allow for a smoother transition because the process already is working under manual control and tuning parameters and control relationships may be known. Also, if any problems develop with the control system, control can easily be switched back to manual.

Having a working process does require that the control system staff be trained adequately and ready to take over from the contractor when the time comes. This will help instill confidence in the system in the plant O&M staff.

Organizational Possibilities

As the control system staff is usually relatively small, there are not many possibilities for organizational structure. The most important requirement of the organization is that the system manager have overall responsibility for the O&M of the control system. The programmers and maintenance technicians then should report to the system manager.

Traditional Organization

The traditional organization of the control system staff has the system manager in charge of the control system and the programming and maintenance staff in two parallel branches reporting to the system manager. The system manager should have a fairly responsible and senior position within the organizational structure of the plant itself to have the necessary authority and influence to obtain budgetary and personnel considerations.

Process Engineer Structure

In this structure, typically a smaller plant, the process engineer is also the system manager. This makes good sense because the process engineer knows the most about the process and probably was involved intimately in the justification and acquisition of the computer system. If the process engineer does not have the luxury of software and hardware maintenance people assigned directly (which is preferred), they probably will come from the O&M plant staff. The software people could come from the data processing staff. This situation is not as desirable because payroll problems tend to take on more importance than process problems. In addition, it is very difficult to find a programmer who is competent in both business data processing and real-time process control.

Importance of the Structure

Some reasons for having a defined organizational structure have been discussed already. There are others that are equally important but sometimes not as obvious. It is not the intent of this chapter to enter into a discussion of personnel management practices but a few considerations should be noted.

No one person should report to more than one supervisor and, conversely, the supervisor must be accountable for the actions of subordinates. These two requirements dictate a defined organizational structure. The efficiency of an organization will increase as the work is directed toward the goals and objectives of the organization.

The employee who can identify with, and feel a part of, an organization, will be more motivated to contribute to its success. When planning for a new organization, those people who will become involved in the organization should be involved in the detailed planning. They will have had a part in building something new and likely will take more pride in it.

PERSONNEL QUALIFICATIONS

The capabilities and knowledge of the members of the control system staff are what make the difference in a smoothly operating plant. Attitude and initiative are also important. However, attitude and initiative only determine the extent to which the person's capabilities are used. In the description that follows, no attempt has been made to define complete position descriptions. Rather, a few key capabilities and strengths have been noted to form the basis for position descriptions unique to the plant organizational needs. The usual attributes of working well with other people, showing individual initiative, using time efficiently and taking direction from others also are not covered here but are nonetheless important.

Operators

Operators must have a thorough working knowledge of the plant processes, including chemical and physical reactions, equipment operating controls and parameters, structure capacities, electrical and mechanical ratings, and piping and vessel capacities. Operators must be intimately familiar with emergency operating procedures and be able to respond quickly to off-normal situations.

Operators must be intimately familiar with all the control and monitoring functions of the control system and how they relate to the plant instrumentation and control equipment. For example, a skilled operator should be able to make an immediate evaluation of a malfunctioning instrument and determine what, if any, corrective action should take place in the control of the process.

The operator's knowledge of, and familiarity with, the plant and the control system will vary with the amount of responsibility assigned. At the same time, operators should be given an opportunity to advance by learning and gaining more experience about the various parts of the system.

The senior operator must be able to read and interpret electrical schematic and wiring diagrams, ladder diagrams, and cable schedules. The senior operator must be able to read and interpret logic drawings, piping and instrument diagrams, mechanical general arrangement and piping drawings, and structural details.

The operator must be able to read and understand O&M manuals for the plant process, instrumentation, control and monitoring equipment, and be able to evaluate operational problems quickly.

Operators should have a minimum of high school graduation. The more senior operators should have demonstrated their initiative by attending supplementary trade school courses. They should attend supplier training courses whenever they become available. The number of years of experience will vary but should approach 3-5 for senior operators.

Digital Technicians

Digital technicians must have a thorough working knowledge of electronics, integrated circuits, digital logic, memories, and all the equipment and devices associated with digital computers. They must have a working familiarity with software programming and be able to write and apply small but complex diagnostic routines to check hardware operation. They must be able to operate and interpret results from test equipment such as oscilloscopes, digital logic testers, multimeters and circuit tracing instruments.

The digital technician must be able to read, understand and apply hardware maintenance manuals, logic drawings, electrical schematic and wiring diagrams, cable schedules, connector diagrams, and mechanical disassembly and reassembly drawings. Additional skills include knowledge and proper application of precision tools.

The digital technician's formal education should include high school plus two years of trade school in electronics. Military service schools also could be applicable. Digital technicians should have had training courses given by manufacturers and suppliers of digital computer equipment. In smaller plants, where the digital technician also may be the instrument technician, a knowledge of instruments and control equipment will be required.

Instrument Technician

Instrument technicians must have a thorough working knowledge of measurement and control theory and be able to apply the theory to the repair, calibration and tuning of instruments and control devices. This is generally highly skilled work involving precision tools and procedures to repair, adjust, clean, test and replace electrical, hydraulic, pneumatic, electronic and mechanical instrumentation and control equipment.

The instrument technician must have a working familiarity with the plant process piping and equipment and be able to analyze and diagnose instrument problems by relating to other equipment operating parameters. The instrument technician must have knowledge of, and respect for, operating processes and temperatures and be thoroughly versed in the safety aspects of working with pressures, temperatures, rotating machinery and electrical equipment.

The instrument technician must be familiar with, and able to use correctly, the precision tools associated with instrument disassembly, repair, replacement and reassembly. He must be able to use precision electrical and mechanical testing and measuring equipment and to fabricate new mechanical parts and repair damaged ones.

The instrument technician must be able to read, understand and apply maintenance and service instructions and manuals, including wiring diagrams, logic drawings, electrical schematics, mechanical assembly drawings, piping and instrument drawings, cable schedules, piping details, and structural details. The instrument technician's education must include high school and a trade or military service school. Maintenance and service courses from the various manufacturers and suppliers of instruments will be required.

Software Programmers

Software programmers must first have a thorough working knowledge of the theory and techniques of digital computers, including digital logic, numbering systems, memories and all software systems associated with the particular computer in use. Programmers must have the ability to retain large amounts of logic rules, procedures and sequences and be able to apply them quickly to the analysis and solution of applications problems. Programmers must be able to translate the high-speed timing relationship of digital logic functions to the slower real-time operations of plant equipment, instruments and controls, as well as auxiliary computer equipment.

Programmers must be able to read and interpret software instruction manuals, timing diagrams, truth tables, logic diagrams and wiring diagrams, and be familiar with and understand hardware logic functions. Programmers must be able to read, interpret and understand machine and assembly language coding, as well as higher level language coding, whether done by themselves or by other programmers.

A most important quality of software programmers is an even, patient temperament. Programmers must be able to work at solving software problems sometimes for long, hard, frustrating hours. They also must instill in others the desire to work along with them until the problem is solved because controlling the plant may depend on the solution. Programmers must be able to communicate with, and gain the respect of, the plant and system operators. To understand how their software interacts with the plant equipment, programmers should have at least a basic familiarity with how the plant process equipment and instrumentation and controls work.

The software programmer should be a college graduate in mathematics, computer science or some related engineering field with heavy emphasis on computer science. Specialized training in the computer and software languages of the plant control system will be required.

Data processing programming experience sometimes can be helpful, but the heavy requirements for science and engineering technology make this transition difficult.

System Manager

The system manager is the key person on the control system staff and, therefore, must have good solid training and experience in both management and technical fields. The system manager must be able to take independent action and be able to stick with projects to completion. The system manager must have the necessary skills to lead a group of people under a great deal of pressure from plant management and plant operations and maintenance. A new computer system will be blamed for many problems not of its own doing, and the system manager must be able to instill faith and confidence in the system.

The system manager should have a good working familiarity with computers, computer systems, and the plant instrumentation and control equipment. This person should have a working knowledge of the plant equipment and the processes that the control system is monitoring and controlling. The system manager must be familiar with, and keep up to date with, any applicable regulatory agencies and their rules and regulations.

The system manager should be a college graduate in science or engineering and have 5-10 years of increasingly responsible experience in running a system. However, education is much less important than solid experience coupled with dedication and a burning desire to see the system function to its utmost. The system manager's most important quality is the ability to achieve results through people.

CONTROL SYSTEM STAFF

The system manager often is faced with a catch-22 situation—finding highly qualified and motivated people to operate and maintain the system who will work for low wages and benefits. This, of course, is not unlike any manager's problem, so the job is not unique.

Hiring and Promoting from Within

The best place to obtain motivated employees is from within existing company staff. These people are trained in company policy and procedures and have made some commitment to work. As the new system takes form, these people can be involved, thereby making them feel a part of the project. This is a highly motivating technique and will produce considerable job satisfaction.

The system manager will need to survey the available staff, determine who would be affected by the new system, and try to match qualifications and interest level with the new positions. However, one should not prejudge employees too much by what they have shown in the past; it might require a new challenge to reveal their talents.

All of the existing plant O&M people should be exposed to some preliminary training about the new system. This might be a good time to evaluate the interest level of the existing employees. Those who perform well and expend some extra effort during the initial training likely will become the most productive control system staff members. This initial exposure to the new system should not be too detailed; rather, its main purpose should be to acquaint the students with the type of equipment and how the plant processes and equipment will be controlled.

SUCCESS STEP 4—STAFFING

The people who will be selected to become part of the system staff should know that promotions and salary increases will reward their improved performance and dedication. Increased job satisfaction from personal growth will provide both long- and short-term benefits.

Some examples of upgrading existing positions are as follows:

- Upgrade instrument technician to computer maintenance technician.
- Upgrade field operators to control room operators.
- Train extra-clever instrument technicians or maintenance electricians to become software programmers.
- Retrain data processing programmers to become real-time programmers.
- Upgrade extra-clever chief operators or plant engineering personnel to become future system managers.

Other plants within an organization are also sources of qualified personnel, provided that all sources within an existing plant are exhausted. Of course, a new plant will be required to find personnel elsewhere. The new positions can be very effective incentives to increased performance by employees throughout the organization. The new control system should be presented as a challenge and an opportunity for increased responsibility and reward, rather than as a threat to existing positions. This will require careful planning and presentation by the system manager and plant management.

Source of Staff

The decision to purchase and install a new control system should be accompanied by a decision to hire the proper staff to maintain and operate it. Both decisions are required to ensure its success.

The first staff member must be the system manager—the key person. The system manager should be on staff to participate during the discussions and decision to implement the control system. Once the decision has been made, the system manager's primary responsibility is to prepare both the plant and the plant management and staff to accept the system. Some of the manager's activities should include trips to other plants to review first-hand installations similar to what is planned. Personal contact with other system managers to discuss potential problem areas will be an invaluable aid to precluding similar problems with the new system.

The system manager, along with key supervisory personnel as they are obtained, should be sent to those manufacturers' and suppliers' facilities who have been selected as qualified suppliers for the new system.

The system manager will be an invaluable aid to the design engineers.

The system manager will be able to provide first-hand information to the designer about how the plant works and what some of its peculiarities might be. If the plant is existing, potential replacement or additions to existing instruments and control equipment can be identified by the new system staff. Finally, the staff should be involved with review of the design of the system.

Participation of the control system staff is essential during the construction, installation, testing and startup of the system. Problems solved during these activities will be invaluable to the staff in understanding and appreciating the system's capabilities. No amount of book training and lectures can replace actual on-the-job training received by solving real-world problems.

A balance must be maintained during the periods of low activity and the size of the staff. Hiring too many too soon will result in nothing to do, boredom and a waste of hard-to-come-by dollars. A good way to utilize existing personnel who are being upgraded is to split their time between existing and new activities. This also provides continuity to their previous positions.

The goal is to have a trained and capable staff which is fully qualified and ready to assume responsibilities for the new system when the system is ready for the staff. Some activities which the control system staff can perform during construction, installation, testing and startup are as follows:

- Assist the supplier in identifying and solving interface problems.
- Receive hardware and software training from the control system supplier.
- Assist in writing procedures and policy manuals for the new control systems staff.
- Assist in reviewing final documentation and O&M manuals provided by the supplier.
- Assist in reviewing submittal shop drawings.
- Assist other plant personnel in modifying their job functions as affected by the control system.
- Monitor plant construction to be intimately familiar with the plant.
- Assist supplier in testing I/O circuits to the control system.
- Assist supplier in testing and demonstrating system control strategies.
- Operate plant processes manually during installation and startup of the control system.

CONSIDERATIONS FOR ABNORMAL SITUATIONS

The system manager and the system staff must be prepared for abnormal situations that may occur during the day-to-day and long-term operation of the system. This preparation will include cross-training of personnel, preparation and practice of emergency operating plans and procedures, and locating and procuring agreements for outside equipment and services.

Labor Disputes

Labor disputes are difficult to deal with as personalities of fellow staff members can be altered by the issues under discussion. Plans must be made not only to protect the system but also to maintain control over the plant processes.

Software is especially vulnerable before, and during, a labor dispute. Therefore, arrangements must be made to have a secure place to store a complete set of system software. This storage location must be accessible only to nonunion personnel, as well as being safe from fire, theft and stray electromagnetic fields. Backup software also is required in case of catastrophic failure with the running system. The supplier of the system often will maintain a separate set of documentation or records of the system in secure archives. While this set may not be always up-to-date, it would be far better than starting over.

Backup Panels

In some plants it may be necessary to have backup control panels situated throughout the plant in cases of emergency. If the central control system or the communications from central to the remotes should fail, manual operation from the backup panels will be required. The extent and complexity of the backup panels can vary from a single pushbutton control station to an elaborate control panel with indicators, recorders, controllers, manual loading stations and switches. Regardless of the complexity, some form of communications between the various panels and the central control room must be provided. This will be especially critical during a labor dispute because the availability of operations personnel may be severely limited. Procedures must be established and be familiar to nonunion supervisory personnel.

250   PROCESS CONTROL COMPUTER SYSTEMS

Contingency Plans

The difference between a plant that continues to operate and one that does not under emergency conditions is the existence of well-conceived and practical contingency plans. As school children have disaster drills, so the system staff must be made to perform under similarly demanding situations. This requires regularly scheduled training and refresher courses for new and existing staff, respectively.

Outside services that can be mobilized quickly when needed should be considered. Thought also should be given to the demands on them if the same causative conditions are widespread. Some examples of outside services are instrument repair, programming, computer hardware maintenance, radio communication, food services, sleeping accommodations and cleanup labor.

SUMMARY

Throughout this chapter, many statements have been made about actions that should or should not be taken. Some are obvious, others are not. The following list will point out the major intent of the chapter:

Do's

- Do obtain a qualified, competent and dedicated key person—the system manager.
- Do establish and maintain clear policies and procedures.
- Do emphasize personnel training.
- Do train control system staff in plant processes.
- Do assign responsibility for the system to one person and hold that person accountable for the system and staff.
- Do maintain the hardware and software.
- Do challenge the staff to excel and reward them for it.
- Do keep a backup software system.
- Do balance personnel and hardware.
- Do rotate personnel as much as possible.
- Do obtain staff early in the development of the system.
- Do establish direct lines of authority and responsibility.
- Do plan for contingencies.
- Do encourage daily communication between operations and maintenance personnel.
- Do practice for disaster situations.

- Do provide orientation training for the entire plant staff in the concepts of computers, controls and automation.

Don'ts

- Don't let the system or the staff stagnate.
- Don't forget the vendor.
- Don't prejudge existing employees when recruiting new system staff.
- Don't assume the computer will solve all problems; the computer can only implement solutions.
- Don't assume that "turnkey" means troublefree or unchanging. The system will need to be updated, changed and, most importantly, managed.

CHAPTER 11

SUCCESS STEP 5 – MANAGEMENT OF PROCESS CONTROL AND PROCESS OPTIMIZATION

Robert G. Skrentner, PE

INTRODUCTION

The purpose of this chapter is to address management responsibilities, general management approaches and details of both the day-to-day and long-term management of the control system.

Previous chapters have discussed the success steps to plan, design, implement and staff a digital control system. The objective of the preceding steps is a successful control system. Many control systems were implemented following similar steps and, on startup, were relatively successful. However, no matter how well planned, designed, installed or staffed, no control system will be successful without a management commitment to make it work over its expected life.

In this chapter are a number of rhetorical questions along with examples of successes and failures. Every plant has its own individual needs and problems. Every manager has his/her own style of management. It would be presumptuous to dictate management practices. Therefore, examples and questions are used as a means to stimulate thought.

In addition, some specific recommendations are given. These deal with the development of a resource person to assist in managing the control system and with the development of procedural guides for the day-to-day and long-term management and operation of the control system.

THE MANAGEMENT PROCESS

Management is the process of accomplishing tasks with and through people by guiding and motivating their efforts toward common objectives [1]. The management aspects of all supervisory jobs are the same regardless of the technical content or level within the organization. The management functions of planning, organizing, staffing, directing and controlling are common to all supervisors and managers.

Within these five areas of management functions is the process of solving problems, making decisions and implementing plans [2, 3]. Problem-solving involves finding the cause of a problem. Problems may include poor product quality, excessive inefficiency or wastage, excessive manpower utilization or poor employee morale. Decision-making is the process of deciding what to do to eliminate the cause of a problem. Once a decision is reached, it must be put into practice by implementing an organized plan.

Knowledge of Content and Delegation Downward

A manager is concerned with the overall product quality and efficiency of the operation. The manager may not know every detail of the operation, but is responsible for its success or failure.

Authority and responsibility must be delegated to provide the manager with the information necessary to perform effectively, without being overwhelmed by details [4]. As tasks are delegated down through the plant organization, each level knows more about particular details of a part of the plant operation. For example, the operations superintendent will know more about the operation than the manager but less about other plant functions, such as personnel acquisition. The process operator will be thoroughly familiar with his particular process area of concern but may not be familiar with other processes within the plant. Thus, the higher in the chain of command one rises, the more one must rely on subordinates to provide information to assist upper management in problem-solving, decision-making and implementing plans [5].

The Key Person

One cannot be expected to know all of the details of the control system. Therefore, one must develop a resource to provide management information. This resource is a person(s) who has the time and inclination to

PROCESS CONTROL & OPTIMIZATION MANAGEMENT 255

become thoroughly familiar with the control system—the key person [6].

In most plants, it is unlikely that this person will be available at the start of the control system project. Three alternatives are available for acquiring the personnel resource: (1) selection and training from within the organization; (2) transfer of skilled personnel from other divisions within the organization; and (3) the hiring of outside expertise. Regarding hiring from without, often control system managers are not readily available and those who are available will command a relatively high salary. Other divisions within the organization will strongly resist transfer of their key personnel. Therefore, it is likely that someone will have to be trained to fill the position. It may take two to four years for this individual to acquire all the necessary skills.

One should start with an intermediate-level individual. This person may have an electrical, chemical, or civil engineering background, be an electrical technician or even have an operating background. The main criterion is that the person understand conceptually how various process parameters interact. The individual must be able to visualize how the process works and how changes will impact it. The individual should be able to relate to operators, maintenance personnel and management and should be capable of assuming a management position.

Regardless of who is selected, that person will require additional training that is geared to the person's background. The best method of training is a combination of on-the-job experience interspersed with formal training, as the need is perceived. The best time to begin training is during the planning or design stage of a project. The individual should become familiar with the rationale behind the design and not just the end product of the rationale. The person should carefully monitor construction activities with the objective of learning the details of the system, not the actual construction management. If development work is being done at a vendor's factory, this person should be allowed frequent trips to learn the details of the system and the vendor's philosophical approach to the implementation. This person must be available for startup and checkout and should be the representative for the operations group. This is a "baptism by fire" type of training but it is effective. Much assistance in this training can be gained from the design engineers and the vendor's applications engineers and programmers.

The key person is ultimately being trained to become a manager. As such, the individual does not have to know every detail of the system. The key person must learn enough to make intelligent recommendations as to additional personnel needs, system change management and other duties (discussed later in this chapter). For a small system, this person may perform both the management and detailed day-to-day assistance, such as operator training, programming or control changes, system analysis and recordkeeping and other more detailed functions. For a large control system, this person may be almost exclusively management oriented.

GENERAL MANAGEMENT PRACTICES

Two extremes in management practice are crisis management (reactive), in which the manager is continually reacting to events, and controlled management (directive), in which the manager causes events to occur based on management actions. In the former, the manager is like a fire fighter, continually putting out "fires" such as poor product quality, plant inefficiencies, low employee morale, excessive accidents and a host of other problems.

On the other hand, some managers seem to have very few operational or employee problems. Many times one hears from plant personnel that the plant runs itself; however, this actually does not happen. People run the plant. One usually finds that the management and staff of this plant are fire preventers rather than fire fighters. The management and staff are in control and the plant success reflects this.

Most of a manager's activities fall somewhere between the two extremes. Everything may be under control when the manager leaves on Friday, but often a crisis has arisen by Monday.

Crisis Management — What Not To Do

With the advent of computer control systems, many managers were faced with a new dilemma. Computer control systems provided centralized plant monitoring and control functions as their prime objective. In addition, most systems could produce operation and maintenance (O&M) reports, store data for long-term analysis, accept inputs from laboratory test results and numerous other capabilities. Many of these systems crossed lines of authority and responsibility. Managers were faced with a computer they did not fully understand, a computer that could perform many functions that previously had been performed by various groups within the organization. In many cases, the manager had no one to turn to who could provide information required for proper execution of the manager's functions. The manager was faced with a crisis: "What do I do with this computer system?" Of course, managers are creative individuals and a number of successful tactics were developed to avoid addressing the crisis.

Defer Decisions

The first tactic applied by managers is to defer taking over the operation of the computer system. A common ploy is to tell the contractor or engineer that operations personnel were not fully comfortable with the adequacy of the equipment. "Let's wait and see how it performs before rushing into the decision to accept it." This tactic is quite valid if one is willing to pay the computer supplier or contractor extended job costs or subject oneself to some type of litigation.

Another ploy to defer takeover is the "What if..." approach. "What if the computer fails?" "What if the contractor cannot or does not maintain the system?" "What if the documentation is not adequate?" "What if the operators cannot figure out how to use the system?" The intent is to ensure a sufficient stream of questions to keep the computer supplier so busy that he does not have time to insist on acceptance of the system. This ploy is only good for a limited time period.

Failing in the above attempts, the creative manager can transfer blame.

Don't Answer Questions

Once the system is accepted and is being used to operate the plant, many potentially embarrassing questions can crop up. Some typical examples include: "How do we know it's working properly?" "What do we do if it fails?" "What do we do with all the paper?" "Where in our union contract does it describe what the control room operators duties are?" "How can we control the plant when maintenance never fixes anything?" "Why does the field operator always want to place the controls in the local mode instead of the computer mode?"

The fire fighting manager never considers answers to these types of questions. More likely, if this is the first computer control system, he/she never realized what types of questions might be asked in the first place. It is difficult to plan for startup if one is not sure what is involved in a computer system.

There are other creative management approaches to the above. The best is to defer answering or addressing the questions until the problem has been studied.

Delegation

At what point does the operations group begin to hire or train personnel to take over the system? It is hard to justify hiring computer control room operators and process control engineer/programmer type personnel a year or two in advance of delivery of the control system. In

258 PROCESS CONTROL COMPUTER SYSTEMS

many cases, budgetary or political constraints have made the acquisition of personnel on a timely basis very difficult.

Process control engineer/programmer personnel take time to train. Without some long-range planning, the computer system will arrive with no one from the operations group who is thoroughly familiar with it.

Faced with this situation, the creative manager finds a staff member who has heard about or seen a process control system. This person then becomes the expert and all questions can be delegated.

Neglect the System

As a last resort, the manager can neglect the system, which has occurred in many installations. The use of the system slowly decreases and, in time, the plant reverts to a predominantly manual mode of control. The computer system may be used for monitoring only or may be abandoned entirely. Of course, the creative manager has developed and documented various reasons why this has occurred. Because few, if any, of the upper management personnel can refute these reasons, the manager is protected.

System Management

It is probably safe to assume that all managers have applied some of the preceding tactics at one time or another. They can be useful in certain situations. The key question is: Are you actually buying a little time or are you creating another fire?

A digital control system may be thought of as the "eternal triangle." At the base of the triangle are the people who will operate and maintain the system. One side of the triangle is the field equipment, including panels, instrumentation, control devices, and process tankage and piping. The other side is the digital control system, including the computer hardware and programming (Figure 1).

System management may be thought of as the systematic direction of people to bring about the desired results by utilizing the control system components in the most cost-effective manner. As shown in Figure 1, management is in the middle of the triangle. System management is the force that holds the triangle together.

Level of System Management Effort

The degree of management effort required is in proportion to the size and complexity of the control system. Small, simple control systems have

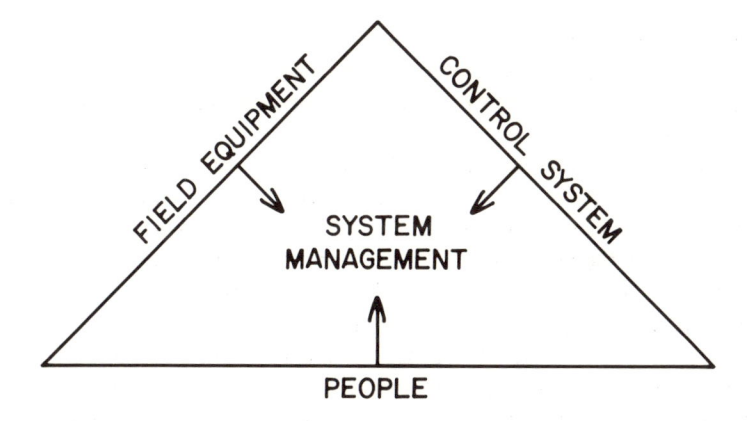

Figure 1. The "eternal triangle."

fewer people, less field equipment and fewer controls. In general, these systems create the fewest management problems. It is relatively easy for a manager to know what is going on throughout the plant at any given time.

On the other hand, large, multifunction systems tend to have the most potential management problems. In these systems, numerous users are making demands on the time of the computer facilities. Groups within the plant organization must be coordinated to ensure that all of the interrelated functions performed by the computer system and the O&M and lab personnel are carried out in the most cost-effective manner.

The organizational structure of the plant also will affect the level of management. In some plants, the installation and startup of the process and digital control system is closely supervised by the operating division. This helps the O&M personnel to learn the idiosyncrasies of the system before actual process operation begins. Many coordination and management problems can be resolved during this period. This is the best organization and, if at all possible, should be implemented.

In many organizations, an engineering group or a consulting engineer working with the contractor or supplier is responsible for installation and startup of the process control. In this case, the operations personnel may not have the time or responsibility to learn the details of the control system. The operational takeover becomes the actual shakedown period to resolve both operational and management problems. As plant productivity must be maintained during this period, success is much more difficult to achieve. In this case, it is critical that the key person be involved heavily as the representative of the operations group providing the liaison between construction and operations.

Other organizational considerations include the impact of unions and past management practice, corporate or upper management-level commitment to success of the operation and the commitment of lower-level personnel to make the system work.

One major point to be made is based on the early experiences of the data processing industry. People who must interact with the computer tend to fear it. Part of this results from lack of familiarity with the computer and how they will use it. If the operator learns that a control panel is being replaced with a new panel, the operator probably will offer little complaint. If the operator is told the control panel is being replaced with a computer, many concerns will be expressed. It is crucial that operating personnel be informed early on in the project that they are not being replaced by a computer and that the system is not intended to make their lives more difficult. Many data processing installations have had numerous startup difficulties because computer operators were not briefed adequately on the purpose of the equipment.

In one case, the plant manager and his operating staff were briefed on the function of the computer control system prior to its delivery. Several briefings were held on Saturdays when all operators could attend. Topics included what the computer would control; how the operators would use the computer; what kind of startup problems might occur; vendor training schedules and topics; and how the field operator and control room operator could work as a team. Many question-and-answer sessions took place.

These briefings allayed operator fears of the system and provided a foundation for subsequent vendor training and operator use of the system. The operating staff was motivated to make the system work.

Unfortunately, several events practically destroyed this effort and severely demoralized the operating staff. Of the 47 computer control strategies, only two on/off control strategies worked on startup because few of the field instruments and control devices worked properly. The plant had been started up in a manual mode, and there had been little checkout of local backup control panels, instruments and control devices. The construction engineering group did not consider automatic controls to be an important item and placed these inoperable items on a punchlist.

The situation deteriorated further. The contractor did not fix many of these items. When a few analog control loops were made operable, it was found that the computer did not work properly for analog control. As the plant was meeting production goals (although not very efficiently), the main office management gave repairs a low priority. The construction group left the site, operators left the control room and the plant manager was left with a lot of junk.

Fortunately, the plant manager had the foresight to anticipate these problems. He had included funds in his operating budget to solve these problems. He was able to force the contractor to fix the computer. Working with the operators and part-time instrument repair group on loan from another plant, he prioritized plant areas for repair. After about 18 months, he was able to get 30 control strategies operable. Many other equipment items can be controlled manually through the computer.

Operators are using the computer control, and morale has improved dramatically. The manager was not willing to accept an inoperable control system. He had to overcome many obstacles due to upper management and other divisions within the organization, as well as some of the more typical construction problems.

PROCESS CONTROL MANAGEMENT

System Functions

The main function of a computer control system is to aid in monitoring and controlling the process. Other possible functions include the storage and retrieval of operating data, manual data such as laboratory data or quality control data, maintenance data or other types of nonprocess control data. The computer may monitor equipment run times to produce preventive maintenance reports. It may be used for personnel record-keeping, statistical analysis of historical data or other functions limited only by the creativity of the user.

System Users

People are at the base of the control triangle. Successful systems depend heavily on the users as the base of the control system. Users include:

1. operators who monitor and control the process through the control system;
2. managers who can use the system to monitor the operating efficiency of the process;
3. lab personnel who enter the results of various quality control tests to be used by both the operators and managers;
4. process engineers who analyze both the process operational efficiency and the control system efficiency;
5. programming personnel who implement changes to the system; and

6. maintenance personnel who may use the system to assist in diagnosing and correcting problems with either the process or the control system.

For small control systems, the number of users is limited to one or two operators per shift and perhaps a process engineer/programmer or maintenance technician. For larger, centralized control systems, the number of users may increase substantially. There may be four or more operators per shift. There may be several process engineers who are responsible for individual plant areas. There will be field operators, various shift supervisors, a number of maintenance personnel and perhaps several levels of management who look at the results of the plant operation.

Prior to discussing some of the management functions, an amplification of some of the various duties of personnel will point out some of the areas for potential management problems.

Operator Use

The first responsibility of the operator is to monitor and control the process. This is done in three major ways:

1. Observe process operation. Most control systems provide at least three ways for the operators to monitor the process. They can use the video display terminal (VDT) displays, which may take the form of bar charts or graphic-type displays. They can trend data on strip charts or on the VDT displays. An alarm/event logger is provided to log messages to the operator.

Observing the process operation sounds like an easy task. However, there are many subtleties involved. All operational data cannot be displayed on one VDT screen. Therefore, in what order should the operator observe the process? How does the operator verify the accuracy of what he is observing? How does he know that a flow value, for example, is correct? How does he know that a critical alarm sensor is working?

2. Make control adjustments. This involves making adjustments to the plant operation. It may include changing flow setpoints, starting up or shutting down processes or pieces of equipment, and responding to equipment failures or process alarm conditions. Again, this task may not be as easy as it sounds. Who authorizes the operator to make changes? What is the operator's level of authority and responsibility? Does the operator know how to respond to major process failures or alarm conditions? Who is notified in the event of problems?

3. Coordinate field O&M activities. The rationale for installation of a centralized control system is to concentrate operating data in one spot and to coordinate the operation of the plant or process. In some plants, this seemed good in theory but ended up poor in practice. In many cases,

field operators would take equipment off of control and run it manually from their area control station for no apparent reason. Maintenance personnel would take a device out of service without notifying the central control room operators. Many times when maintenance was completed, the device would be left in local control or placed back into computer control without notification of the central operators. Management at these plants could not tell at any given time what was under control and what was not.

4. Other responsibilities may include printing plant operating reports on a shift — daily, weekly, monthly or annual basis; writing maintenance orders, maintaining a shift log book or diary; performing custodial maintenance, such as changing ribbons, paper or strip charts, and cleaning VDT screen faces and keyboards; and assisting other personnel in contingent control testing and maintenance work.

Management Use

Managerial use of the control system is usually limited to the observance of the process performance. This usually takes the form of reviewing operating operating reports and the alarm/event summary reports. A manager may use the VDT displays to observe the current operational status of the plant.

What reports should the manager review? When and how often should they be reviewed? What data are important? How will possible operating problems be noted and corrected? What feedback mechanism to the operators will be followed?

Laboratory Uses

In many installations the lab use is confined to the entry of data to be used in various operating reports. This may be done through a remote VDT located in the lab or through terminals in the control room. In a few installations, the lab has much greater input into the operation of the process and may make extensive use of data processing capabilities of the computer system to perform various statistical analyses of the lab and process data.

What data will be entered, when and by whom? What will be the use of the data? What is the relationship between operators and lab staff?

Process Engineering Uses

Process engineers are interested in ways to improve process performance. As such, they typically observe trends in the operation or analyze historical process data. In addition, they may make changes to the control system and perform the testing necessary to verify proper process control operation.

What is the relationship between process engineers and the operators? Who has higher authority and priority of use? How is modification and testing coordinated with operations? How is historical data retrieved and used by the process engineer?

Programmer Uses

If the installation is large enough to justify a programmer or programming staff, their duties typically are related to enhancements to the system and indirectly related to the control of the process, or are related to using the system for data processing-type activities.

How are enhancements made and tested? What are potential adverse impacts on the process and how will these be minimized? How will the system be restored to its former state in case of programming errors?

Maintenance Uses

Maintenance uses of the control systems vary from limited use, such as fixing broken components, to extensive use of the system to assist in isolating field equipment in need of repair, to monitoring run times for preventive maintenance scheduling, and for maintenance recordkeeping purposes.

When and how is preventive or demand maintenance performed on control room equipment? Who has sufficient skill to effect repairs? How is field maintenance coordinated with operations? Who updates preventive maintenance data stored in the computer? When? What is done with preventive maintenance reports?

Process Control Management Concerns

The manager must be concerned with the day-to-day operation and the planning for, and implementing of, change.

Day-to-Day Operation

The day-to-day operation must consider actions to be taken under both normal and abnormal operating conditions. Lines of authority and responsibility, priority of use, control room rules, interactions with unions and the daily recordkeeping activities are other concerns that must be addressed.

One must wonder who is really in charge at some plants. There seems to be no clear distinction of the authority and responsibility of the various personnel. The control room operators think they are in charge of the operation. However, maintenance personnel seem to feel free to take equipment out of computer control at any given time. Notification of the central control staff is of secondary importance. The same is true for the various field operating personnel. Many times, devices are taken off central control with little apparent reason. Process engineers may be in the control room asking operators to make process changes for test purposes, which conflict with operational parameters that may have been set by the plant manager or shift supervisor. The operators in the control room are not sure of their level of authority in directing field actions of the maintenance or operating staff. If no one is really in charge, then no one can be blamed when process upsets occur. This is very convenient for the subordinate staff, who rarely worry about the results of the actions on the process.

From the management viewpoint, it can be very frustrating to try to determine what caused an operational problem when no one seems to be in charge. The author has visited a number of plants and asked questions about who directs changes to the process and what happens in the event of various process failures or abnormal operating conditions. The answers in many cases are shocking. No one really seems to know what to do.

Priority of use and control room rules seem to be another area of concern. Who can enter the control room and what can they do in terms of control actions? Are there rules about smoking (a real menace to some types of computer components), eating or drinking in the control room? What does an operator do if he needs to go to the bathroom and no one is available to take over his duties?

When faced with the installation of a central control room, most unions demand that job descriptions be modified for operators because of an increase in responsibility due to centralized operations. Of course, this is also tied to a pay increase. Managers must be prepared to deal with this situation. It could be that the control room operator's duties require higher levels of authority and responsibility. Appropriate pay increases would be in order. It also may be that their duties have not changed significantly from their former work.

Another concern of management should be to monitor the number of grievances filed by the union on behalf of the control room operators. Frustration, boredom, insecurity, responsibility/authority conflicts, conflicting directives and unequal work division among shifts can lead to increases in the number of grievances.

In many cases, the cause of the grievance has little to do with the words stated in the grievance. In one case, grievances were filed about the control room rules that had been imposed. When a curtain was installed over the huge glass viewing windows that had allowed plant visitors and other personnel to stare into the control room, most of the grievances disappeared. The operators no longer felt as if they were working in a fish bowl. This example illustrates the importance of finding the true cause of a problem. In this case, the problem was excessive grievances being filed. The cause was a lack of privacy for the operators, not the rules imposed.

Change Control

It is a mistake to assume that everyone will be pleased with the operation of a control system or that a control system is "cast in concrete." Control systems are meant to reflect the latest operational philosophy and most current plant control objectives. As such, the control systems will change and evolve as experience is gained, as the process itself is modified or as control deficiencies are noted. This topic will be addressed in more detail later in this chapter.

Management Solutions

There is only one management approach to digital control systems or centralized control systems — CLEAR, CONCISE, WRITTEN PROCEDURES [7]. Ideally, these procedures should be developed before the control system arrives onsite and before most of the operating staff is assigned.

There are many methods for developing these procedures. In an existing plant, one may wish to follow established procedures and current documentation techniques. If existing guidelines are not available, a number of industrial engineering techniques have been developed that can be applied to the control system [8, 9]. The remainder of this section lists items that should be addressed by the procedures.

A major item is the establishment of the lines of authority and responsibility, which must clearly identify the functions of each level within the organization. They should include the following:

1. Who is responsible for setting operational parameters?
2. Who may change the parameters and what is the mechanism for change?

PROCESS CONTROL & OPTIMIZATION MANAGEMENT

3. Under what conditions may control be removed from central and what lines of communication are available to do it?
4. What specifically are the operators allowed and not allowed to do?
5. Who has the authority to allow equipment to be removed from service and under what conditions?
6. Who is responsible to coordinate field operation and maintenance activities?
7. What are the duties of each shift?
8. How is information passed up and down the organization?
9. What forms are used to pass information?
10. What followup and feedback mechanisms are to be followed?

A second major set of written procedures is for the actual operation of the plant. These should address both normal and abnormal operating conditions and should include the following:

1. simple step-by-step procedures for operators to initiate change to the process using the control system;
2. the normal sequence of control actions and the expected results of the actions;
3. what to do when things go wrong, including alternative control actions, notification procedures and any associated cautions;
4. priority of action — what is the critical process?
5. routine process reasonability checking procedures; and
6. backup controls testing.

These procedures should be a summary of O&M manuals. The intent is to have a collection of one-page guidelines for the operators that can be accessed quickly in the event of questions.

Other procedures need to be written for the process engineer/programmers, maintenance personnel, and upper and middle management. All procedures must address the following:

1. Who can take what actions that affect the control system?
2. Under what conditions can they take the actions?
3. What notification procedures are required?
4. What records must be kept? Who receives copies?

The last procedural handbook should be a set of procedures for modifying the above procedures. This should outline the steps to be taken to initiate change, approve the change and implement the change. Included would be how to update old procedures and communicate changes to all affected individuals.

PROCESS CONTROL OPTIMIZATION

> All progress grows out of discontent with the way things are
> D. Kenneth Winebrenner

It is unlikely that everyone will be satisfied with every aspect of a control system, especially a large, complex digital control system. Some dissatisfaction may be due to minor irritants, or system "bugs," as they are known. Operators may not like the way or the order in which information is displayed. The control system may not control the process as efficiently as it could. Subordinates may not be sure of their authority and responsibility as it relates to the control system.

For the purpose of this section, process optimization is defined as any actions taken to ensure that the plant functions at its most efficient level. This includes making sure the existing control is working properly, analyzing the process to note areas where improvements could be made and implementing improvement. Remember, the control system is a triangle of people, field equipment and the control system. Each side is subject to optimization.

Process Reasonability Checks

All control systems, no matter how simple or how complex, depend on measurement of the process operating conditions. To be sure that a process is operating efficiently, some mechanism must be available to ensure that the measurements of the process are correct. Many times operators will observe that a measurement does not look or feel right based on their experience of how a process should act under a given set of circumstances.

Similarly, the computer can be used to assist plant management in obtaining the "feel" of the process. Several methods are available to perform this function. A centralized computer will have the data from many plant measurement devices stored in its memory. These data may be recalled to plot trends similar to strip charts. The operator can use these to observe current trends. In addition, some instruments degrade slowly in performance over a relatively long period of time. Historical data can be trended to observe this degradation. Another method for detecting instrument problems is to perform various balances. For example, if flowmeters are available throughout the plant, the computer can quickly and continuously monitor and sum readings. If the flows do not balance, an alarm can be generated, alerting the operator that some measurements may be incorrect. The computer also may compare current readings against historical norms to detect either off-spec product or potential instrument malfunctions. Another reasonability checking method is to compare two related readings. For example, the flowmeter reading may be compared with a valve position or pump speed. Under most operating conditions there is a correlation between position and flow. Any unusual discrepancies can be quickly flagged to the operator.

These checks do not just happen. They must be managed by establishing procedures for checking readings, recordkeeping, communicating status and management followup.

Control Loop Tuning Reasonability

Many maintenance problems can be caused by improper tuning of control loops. If a loop is tuned improperly, excessive wear may result on the control device. Over time, as devices wear, the characteristics of the device may change and the control loops may have to be retuned to reflect the new operating conditions.

The computer may be used to assist either operators or process engineering personnel to observe the frequency and magnitude of outputs to control devices to detect improper tuning. Some computers have been used to provide self-tuning. Based on accepted norms, the computer will automatically adjust the control loop tuning to provide the most efficient control response. As this type of advanced computer usage is the exception rather than the rule, most checking of loop tuning will be performed by the operators or process engineers, and problems will be reported to the appropriate repair technician.

It is usually easy for operators to detect very poorly tuned loops because there will be abnormal deviations from the desired process setpoints. It is the area of minor tuning adjustments that is most often overlooked. Over the long term, damage may result to equipment from excessively small control adjustments by the automatic controls. Procedures must be established for performing the checks, just as was done for reasonability checks.

Process Analysis

All processes should be evaluated periodically to determine whether the process is operating at peak efficiency. This evaluation may include the number of times and the magnitude of excursions of the actual from the desired product quality, areas in which additional efficiency of operation may result, such as lowered power consumption or decreased chemical usage, and areas in which information could be better displayed to operators to reduce operator errors or improve operator efficiency.

This analysis may take the form of analyzing historical data stored on the computer, observing operator actions in the control room, process experimentation and change, and other similar functions.

Backup Control Testing

Many times, operators become so reliant on the automatic control systems that they cannot manage the process to maximize operational efficiency under the failure conditions when a failure occurs. Backup testing is the procedure used to ensure that the process will continue to operate efficiently under various failure conditions. Various types of failure modes of operation must be created on a regular basis to allow operators to practice under carefully controlled conditions. It is not good practice to wait until the failure occurs and then see how everyone does.

Removing Minor Irritants

No control system is perfect. There always will be some complaints about its operation, the way information is displayed or the way in which it must be operated under various conditions. Digital control systems have the advantage that changes may be made with little, if any, capital cost. Minor irritants, if left unaddressed, can become major operational headaches either in terms of decreasing utilization of the control system or in increasing employee dissatisfaction.

The ease of change of digital systems also can be a disadvantage. Some employees may expect changes to be made based on their personal preferences, rather than for the good of the operation. In addition, changes must be carefully documented and tested, and all affected parties should concur with the change [10].

Controlling Change

In digital systems, change can be one of the most serious potential problem areas. Undocumented changes can be devastating to the operation because generally one cannot see what change has been made.

When a device is removed from a control panel, everyone can see that something was changed. When some parts of a program are removed from a computer, no one may ever know it has occurred.

In one instance, a programmer could do some marvelous things with a computer system but did not believe in documenting his program changes. He was the only one who knew both how to use the program and the details of what it did. This may have been an attempt on his part to add some job security; however, finally his access to the computer system was forbidden until he documented each change.

In another case, the night shift operators did not like the way some information was displayed on the VDT. As this was relatively easy to change, it was not long before changes started being made to the displays. After several weeks of these changes, the computer system failed. It was restored using some backup disks that were about two months old. Needless to say, all the changes were lost.

Subsequently, procedures were established with the night shift operators and the day shift programming staff to document changes and make sure that periodic updates were made to the backup disks. By allowing the operators to make changes under controlled conditions, they gained some ownership of design of the system and tended to support the system and become more comfortable with it.

Impact of Change on Process Operation

A problem in this area came in the testing of a new program to control some pumps. The process engineer decided to go into the field to switch the pumps into computer control. The control panel was in a different building from the pumps. After switching to computer control, the engineer noticed smoke coming from the pump building. When the computer first assumed control, the output signal from the computer to the pump control was at the on/off switchpoint of two pumps. The local pump control had not been checked out thoroughly and was not adjusted properly. The 150-hp motor turned on and off 21 times in less than a minute, finally burning up. The mistake was twofold: this backup mode was not thoroughly tested, and it was assumed it would work the first time.

All control system changes are intended to improve process performance. Unfortunately, the converse may be true. Some managers get around this by forbidding changes to be made as long as the process is operating satisfactorily. Sometimes, their definition of satisfactory leaves a lot to be desired. Managers should not fear change. However, it must be controlled carefully.

Management of Optimization

As in day-to-day management of the operation, the management of optimization requires clear, concise written procedures. The need for documented procedures cannot be stressed enough. The control system will not be successful without them.

An average plant may have 50–100 control loops. How often should the instrument and tuning reasonability be checked? Who is going to do it and what procedures should they follow? How will problems be noted and communicated to the appropriate personnel for corrections? What should be done about the process operation while corrections are being made?

A typical plant may have 5–10 different major processes. How often should the data from the process performance be analyzed? How detailed should the analysis be? Who should do the analysis and to whom should the results be reported? Where is the required data located and how will it be processed?

How often should various failure modes of operation be tested within each process area? Are field operators prepared at all times to assume control from the central control operators if a major failure should occur? What should the central operators do if control is transferred to the field operators?

How can one ensure that changes to the control system do not adversely affect the process during installation, startup and checkout of the control modifications? Who authorizes changes? What resources are required to implement change?

CONCLUSION

Unfortunately, in a text of this type these questions cannot be answered; they can only be posed for consideration. Obviously the answers will depend on the particular plant size and complexity, the importance of tight product quality control, personnel skill levels, available resources and a number of other considerations. One can only stress that these questions should be addressed at the earliest possible time prior to startup and takeover of the operation of the control system.

Management of a digital control system is no different than the management of any interdisciplinary group. One must direct the system to achieve the goals rather than react to events. Since one cannot be expected to know everything about the computer control system, one must acquire the personnel necessary to assist in solving problems, making decisions and implementing plans related to the control system. Clear, concise written procedures lead to the success of any control system.

REFERENCES

1. Haiman, T., and R. L. Hilgert. *Supervision: Concepts and Practices of Management* (Cincinnati, OH: South-Western Publishing Co., 1977), pp. 21, 22.
2. Haiman, T., and R. L. Hilgert. *Supervision: Concepts and Practices of Management* (Cincinnati, OH: South-Western Publishing Co., 1977), pp. 69–78.
3. Kepner, C. H., and B. B. Tregoe. *The Rational Manager* (Princeton, NJ: Kepner-Tregoe, Inc., 1976), pp. 54, 55.
4. Delaney, W. A. *Micromanagement* (New York: Amacom, 1981), Chapter 10.
5. Kepner, C., and B. B. Tregoe. *The Rational Manager* (Princeton, NJ: Kepner-Tregoe, Inc., 1976), p. 39.
6. Skrentner, R. G. "Management of Computer Control Systems—The Key Person Concept," *Poll. Eng.* 13(6):16 (1981).
7. Williams, C. N., and R. G. Skrentner. "Management Planning for Computer Systems," *Poll. Eng.* 13(6):21 (1981).
8. Mallory, C. W., and R. Waller. "Application of Selected Industrial Engineering Techniques to Wastewater Treatment Plants," National Environmental Research Center, EPA Report R2-73-176 (1973).
9. Seiler, E. L., and J. W. Altman. "Guide to the Preparation of Operational Plans for Sewage Treatment Facilities," National Environmental Research Center, EPA Report R2-73-263 (1973).
10. Metzger, P. W. *Managing a Programming Project,* (Englewood Cliffs, NJ: Prentice-Hall, Inc., 1973), pp. 138, 139.

GLOSSARY

access time: (1) the time it takes a computer to locate data or an instruction word in its storage section and transfer it to its arithmetic unit where the required computations are performed; (2) the time it takes to transfer information that has been operated on from the arithmetic unit to the location in storage where the information is stored.

actuator: a mechanism for translating an electronic or pneumatic signal into a corresponding movement or control. (*see also* final control element).

alarms: devices reporting abnormal conditions involving a change in station control; power changes; high, low or reverse flow of water; equipment operating condition; fire; flood; entry; leaks; high water pressure; breakdowns; etc.

algorithm: a prescribed set of rules or procedures for the solution of a problem in a finite number of steps, for example, a statement of an arithmetic procedures for evaluating sinX.

alphanumeric: a character set that contains both letters and numerals and other characters such as punctuation marks.

amplifier: a device that enables an input signal to control power from a source independent of the input signal and thus be capable of delivering an output that bears a relationship to, but is generally greater than, the input signal.

analog: pertaining to representation of numerical quantities by means of continuously variable physical characteristics (contrast with "digital").

analog control: implementation of automatic control loops with analog (pneumatic or electronic) equipment.

analog device: a mechanism that represents numbers by physical quantities, e.g., by lengths (as in a slide rule) or voltage or currents (as in a differential analyzer or a computer of the analog type).

analog signal: a continuously varying representation of a physical quantity, property or condition such as pressure, flow or temperature. The signal may be transmitted as pneumatic, mechanical or electrical energy.

analog-to-digital (A/D): the conversion of analog data to digital data.

analog-to-digital converter (ADC): a device used to convert an analog signal to approximate corresponding digital data.

annunciator: a visual or audible signaling device and the associated circuits used for indication of alarm conditions.

application software: special or customized computer programs whose purpose is to perform some unique functions related to the specific process, e.g., a program to maintain a desired temperature profile in a furnace (*see* system software).

ASCII: American Standard Code for Information Interchange (also known as USASCII).

asynchronous: pertaining to a lack of time coincidence in a set of repeated events; term is applied to a computer to indicate that the execution of one operation is dependent on a signal that the previous operation is completed.

automatic control system: a control system that operates without human intervention.

auxiliary storage: a storage device in addition to the "core" or main storage of the computer. Auxiliary storage is the permanent storage for information; it includes magnetic tapes, cassette tapes, cartridge tapes, hard disks, floppy disks.

availability: the percentage of time that a system or component is ready to function.

background processing: a processing method whereby some computer programs with a low priority are executed only when the computer is not busy with execution of higher priority items.

backup: duplicate data files, redundant equipment or procedures used in the event of failure of a component or storage media.

BASIC (Beginner's All-Purpose Symbolic Instruction Code): an easy-to-learn, high-level language used most frequently with minicomputers, originally developed as an algebraic time-sharing language. More recently, a number of minicomputer manufacturers have extended the limits of BASIC so that it can be used for minicomputer business applications.

batch processing: (1) pertaining to the technique of executing a set of programs, such that each is completed before the next program of the set is started; (2) loosely, the execution of programs serially.

baud rate: a unit of signaling speed, indicating the number of signal changes per second. Most signal schemes have two states representing a bit equal to 1 or 0. In this case bit rate equals baud rate. Some signaling schemes have multiple states. In these, baud rate is less than bit rate.

binary coded decimal (BCD): a decimal notation in which individual decimal digits are represented by a group of binary bits, e.g., in the 8-4-2-1 coded decimal notation, each decimal digit is represented by a group of four binary bits. The number 12 is represented as 0001 0010 for 1 and 2, respectively, whereas in binary notation it is represented as 1100.

bit: (1) abbreviation of binary digit; (2) a single character in a binary number; (3) a single pulse in a group of pulses; (4) a unit of information capacity of a storage device (the capacity in bits is the logarithm to the base two of the number of possible states of the device, related to storage capacity); (5) One binary digit, the smallest piece of information in a computer system; a bit can be either 1 or 0.

bit rate: rate at which binary digits are transmitted over a communications link.

block diagram: a sketch that shows the parts of a control system: computer, peripherals and multiplexers, and how they are linked together.

Boolean: an algebraic system formulated by George Boole for formal operations on true/false logic.

buffer: (1) an internal unit of a computing device which serves as intermediate storage between two storage or data handling operations with different access times or formats; usually to connect an input or output device with the main or internal high-speed storage; (2) an isolating component designed to eliminate the reaction of a driven circuit on the circuits driving it, e.g., a buffer amplifier.

bumpless transfer: transfer from manual to automatic control, or vice versa, without making any changes to the process.

byte: (1) a generic term to indicate a measurable portion of a string of two or more binary digits, e.g., an 8-bit byte; (2) a group of binary digits usually operated on as a unit.

capital: purchased equipment, which is amortized over time.

cascade control: use of two conventional feedback controllers in series such that two loops are formed, one within the other. The output of the controller in the outer loop modifies the setpoint of the controller in the inner loop.

centralized control system: an electronic system using computers and telemetering equipment to collect data, and monitor and control a process at a single location.

central processing unit (CPU): a unit of a computer that includes circuits controlling the interpretation and execution of instructions.

character: one alphanumeric symbol, e.g., letter, figure, number, punctuation or

278 PROCESS CONTROL COMPUTER SYSTEMS

other sign. Characters are usually represented by a code of binary digits, e.g., ASCII.

closed-loop: a signal path formed about a process by a feedback measurement signal (input to a controller) and the signal delivered to the final control element (controller output signal).

COBOL (Common Business Oriented Language): a high-level procedural computer language used for business programming.

common mode rejection (CMR): the ability of a circuit to discriminate against a common mode voltage. CMR may be expressed as a dimensionless ratio, a scalar ratio, or in decibels as 20 times the $\log_{10}$ of that ratio.

common mode voltage: a voltage of the same polarity on both sides of a differential input relative to ground.

compile: to prepare a machine language program from a computer program written in another programming language by making use of the overall logic structure of the program, or generating more than one machine instruction for each symbolic statement, or both, as well as performing the function of an assembler.

computer: (1) a data processor that can perform substantial computation, including numerous arithmetic or logic operations, without intervention by a human operator during the run; (2) a device capable of solving problems by accepting data, performing described operations on the data and supplying the results of these operations. Various types of computers are calculators, digital computers and analog computers.

computer/local (C/L): refers to a switch setting at a local panel. Local setting allows control from local panel only, locking out computer control. Computer setting allows computer control only, locking out local panel controls.

computer programming: preparation of the sequences of instructions to be entered into the computer by an input device.

configuration: a term used when referring to the equipment that will be assembled to work as a unit for a business. It includes the options chosen as well as peripheral devices.

contingency: an action taken in response to abnormal situations.

control diagram: a sketch that shows the parts of a control strategy and how they are linked to the process and to each other.

controllable: a situation where a desired parameter can be manipulated.

controller: a device that operates automatically to regulate a controlled variable by comparing a measurement of the variable with a reference value representing the desired level of operation.

control loop: automatic adjustment of a process component based on a desired value (setpoint) and the current value of the process (controlled) variable.

control mode: specific type of control action such as proportional, integral or derivative.

control philosophy: the general way in which a control system is to be used, the manner of locating control functions and goals of the control system.

control sequence: the normal order of selection of instructions for execution. In some computers, one of the addresses in each instruction specifies the control sequence. In most other computers, the sequence is consecutive except where a transfer occurs.

control system engineering: the specialty branch of professional engineering that requires the education and experience necessary to understand the science of instrumentation and automatic control of dynamic processes, and requires the ability to apply this knowledge to the planning, development, operation and evaluation of systems of control to insure the safety and practical operability of such processes (from California professional registration regulations).

control valve: a final controlling element through which a fluid passes, which adjusts the size of flow passage as directed by a signal from a controller to modify the rate of flow of the fluid.

core: a particular type of main computer memory. Another kind is MOS.

CPU: *see* central processing unit.

data base: a set of data records that are organized in a logical manner so they can easily be accessed and used.

DBMS (Data Base Management System): a set of programs that are used to manipulate and use a data base.

DEAC: *see* digitally emulated analog control.

deadband: (1) a specific range of values within which the incoming signal can be altered without also changing the outgoing response; (2) the range of values of a process variable where no control action is taken. If the process variable exceeds the deadband high or low limits, control action is started.

deadtime: the interval of time between initiation of an input change or stimulus and the start of the resulting observable response.

derivative action: a controller mode that contributes an output proportional to the rate of change of the error.

digital: pertaining to representation of numerical quantities by discrete levels or digits conforming to a prescribed scale of notation.

digitally emulated analog control (DEAC): a control philosophy where the computer (a digital device) duplicates analog controls.

direct acting controller: a controller in which the value of the output signal increases as the value of the input (measured variable) increases.

direct digital control (DDC): a control technique in which a digital computer can be used as the sole controller and its output can be connected directly to the final control element. Used to distinguish from analog control.

discrete: *see* digital.

disk drive: the mechanism within which the disk, storing information used by the computer system, rotates. It is connected to the CPU and contains electronic circuitry that feeds signals from the disk to the computer and back.

distributed control: location of controlling equipment (computer or conventional controllers) at remote locations throughout the process.

distributed processing: use of computers at various locations each of which is tied to a central computer. This allows preliminary processing to be handled by the "distributed" computers and eases the load on the central computer.

disturbance: a change in the operating condition of a process, most commonly a change in input or output loading.

ergonomics: the study of human interaction with machines.

error: the difference between the setpoint reference value and the value of the measured signal.

feedback: the signal in a closed-loop system representing the condition of the controlled variable.

feedback control: control in which a measured variable is compared to its desired value to produce an actuating error signal, which acts on the process to reduce the magnitude of the error.

feedforward control: control in which information concerning one or more conditions that can disturb the controlled variable is converted, outside of any feedback loop, into corrective action to minimize deviations of the controlled variable.

file: a group of related data records usually arranged in some sequence according to a "key" in each record. For instance, a payroll file would contain one record for each employee. It would probably be arranged by the employee number contained in each record.

final control element: the device used to directly change the value of the manipulated variable.

floppy disk: a small flexible inexpensive magnetic disk commonly used for data storage in small computer systems.

flow chart: a diagram that uses decision blocks and statement blocks to outline software programs.

GLOSSARY

foreground processing: a high-priority processing method where real-time control programs and process inputs are given preference (through the use of priority interrupts) over other programs being executed by the computer system (*see* background processing).

FORTRAN (FORmula TRANslation): FORTRAN is a high-level language used primarily for coding mathematical or engineering problems for the computer.

freeze dates: deadlines by which time further changes in design cannot be made.

full duplex: a communications channel with separate circuits for transmission and reception, so that both can occur simultaneously.

graphic: a pictorial representation of a process, shown on a VDT or sometimes on a panel.

half duplex: a communications channel where transmission and reception share the same circuit so that both cannot occur simultaneously.

hand/off/automatic (HOA): refers to a switch setting on a local analog controller. In automatic, the controller will continuously adjust to a setpoint. In manual, controller output can be changed by an operator but the controller will not hold to a setpoint. In the off mode, controller output will hold constant and cannot be changed.

hard coded: a programming practice whereby application-related information, such as point names and ranges, are not represented symbolically within a program such that recompilation is necessary in order to change the information.

hardware: the computer and the peripheral devices attached to it.

hertz (Hz): a unit of frequency equal to one cycle per second.

hierarchy: refers to levels of supervision and control responsibility within the centralized control system.

high-level language: a computer programming language that approaches English in its syntax. Usually easier to learn than a low level-language such as assembly language. BASIC, COBOL and RPG are examples of high level languages.

human factors engineering: the design of machines and environment to accommodate humans.

hybrid control: a combination of analog and digital control.

IDE: *see* interactive data entry.

input/output (I/O): writing to or reading data from a computer's memory or the computer writing to or reading from one of its peripheral devices.

instrumentation: electronic devices that provide monitoring of information, operational control or alarms.

integral action: a controller mode that contributes an output proportional to the integral of the error.

integral time: the time required after a step input is applied for the output of a proportional plus integral mode controller to change by an amount equal to the output due to proportional action alone.

interactive: commonly used to describe a software program that provides give and take between the operator and the machine. The program may ask a question to elicit a response from the operator or present a series of choices from which the operator can select. May also be referred to as "conversational" mode.

interactive data entry: a method of communicating with a computer where the operator is led through a sequence of entry by prompting messages from the computer. It is effective when applied to computer-aided sequence control of process startups or shutdowns.

keyboard: the grouping of keys used for data and command entries.

lateral design: minor modifications of existing equipment or software.

local control: control operations performed at a control panel located near the process, either manually or automatically to back up computer control.

logging: recording values of process variables for later use in trending, report compilation or historical records.

loop gain: the ratio of the change in the return signal to the change in its corresponding error signal at a specified frequency. The gain of the loop elements is frequently measured by opening the loop, with appropriate termination. The gain so measured is often called the open loop gain.

machine language: hexadecimal code the machine has built to be able to interpret directly. Compilers and assemblers translate programmer's code into machine language.

management: the process of accomplishing tasks with and through people by guiding and motivating their efforts toward common objectives.

manipulated variable: the process variable that is changed by the controller to reduce or eliminate error.

manual control: control operations performed directly by a human operator and not by computer control algorithms. Two levels of manual control are possible: (1) local manual—the process is controlled manually from the local panel; and (2) computer manual—the process is controlled manually through computer system interactive VDT displays.

mathematical model: a representation (simulation) of a physical or biological system or a concept, through algebraic equations.

menu: a VDT screen that lists operator options available related to a given area. For example, a graphic menu lists all graphic displays which may be chosen.

microcomputer: generally refers to a digital computer where the central processing unit (CPU) is contained in one large-scale integrated circuit.

minicomputer: generally refers to a digital computer with 8- to 24-bit word length, and where the CPU consists of interconnected integrated circuit chips, although an exact definition does not exist.

modem: a contraction of modulator/demodulator. This is a device that converts digital data into a form suitable for transmission over the communications media. For example, phone line modems convert digital data to audio signals.

monitoring: the information on the conditions of various water control processes, operations, levels and security obtained by electronic devices.

MOS (metal oxide semiconductor): a kind of computer memory in use found in hand calculators. It is cleared when power is turned off. Core, the other kind of memory, retains data even after power is turned off.

MTBF (mean time between failures): a way of gauging the general reliability of a piece of equipment.

multiplexer (MUX): a device that samples input and/or output channels and interleaves signals in frequency or time.

multiplexing: the process of combining several measurements for transmission over a pair of wires or link.

multiprocessing: the apparent simultaneous execution of two or more programs or sequences of instructions by a computer or computer network.

normal mode voltage: a voltage induced across the input terminals of a device.

object language: machine language generated by a program translating system such as a compiler or assembler.

ODC: *see* optimizing digital control.

OEM (original equipment manufacturer): a term commonly used to refer to a computer sales organization that has an arrangement to sell a manufacturer's product. In other industries they would be referred to as dealers or distributors.

on-line: the state of a computer that has master control of the plant processes and all associated real-time functions including data acquisition, alarming, supervisory control and DDC.

on/off control: a system of regulation in which the manipulated variable has only two possible values: on and off.

open-loop: a signal path without feedback.

operating limits: high and low limits set for a process variable. A value of process variable between these limits is considered normal and no control action is taken. When either of the limits are exceeded an alarm or control action is initiated.

operating system: an executive program designed to control the execution of program operations. This includes the dynamic allocation of resources for concurrently running programs on a priority basis.

operational availability demonstration (OAD): a test of the installed system to determine availability and reliability.

operator interface: the means through which an operator accesses the computer system to affect process control actions. Usually this consists of a VDT and keyboard arrangement along with appropriate display software.

operator-process interface (OPI): the hardware and software that allow an operator to monitor and control a process. This includes VDTs, keyboards, displays, printers, annunciator panels and indicators.

optimization: a process whose object is to make one or more variables assume, in the best possible manner, the value best suited to the operation at hand, dependent on the values of certain other variables which may be either predetermined or sensed during the operation.

optimizing digital control: computer control where the analog control capability is enhanced in the computer mode. The objective in ODC is to make the process run better or less expensively than capable with normal analog devices.

parallel transmission: a mode of transmission whereby each bit of a data word is transmitted simultaneously over separate communications circuits.

parameters: the limits or context within which the problem is considered.

parity: one method of error checking in data communications. As a bit string is transmitted, an extra bit is added to make the total number of bits either odd or even (called odd or even parity). The receiving machine checks each bit string for correct parity. If parity is not correct then noise or some other disturbance on the data channel has caused errors in the bit string.

PASCAL: a high-level programming language gaining popularity on microcomputer systems and some minicomputer systems.

peripheral: a device other than the computer itself used in computer processing. Disk drives, tape drives, CRTs and printers are peripheral devices.

piping and instrumentation drawing (P&ID): a schematic drawing of the process showing all liquid flow paths, locations of all sensors and instruments, and location of backup conventional control equipment.

point: a single signal to or from a field device. Points are either analog in; contact (digital) in; modulating (analog) out; or contact (digital) out.

prequalification: a process by which the user/designer compiles a list of vendors qualified to bid a project. Bids are then accepted only from vendors on that list.

primary element: the device that converts a portion of the energy of the variable to be measured to a form suitable for amplification and retransmission by other devices.

priority: level of importance of a program or device.

priority interrupt: the temporary suspension of a program currently being executed in order to execute a program of higher priority. Priority interrupt functions usually include distinguishing the highest priority interrupt active, remembering lower priority interrupts that are active, selectively enabling or disabling priority interrupts, executing a jump instruction to a specific memory location, and storing the program counter register in a specific location.

process: (1) the collective functions performed in and by industrial equipment, exclusive of computer and/or analog control and monitoring equipment; (2) a general term covering such items as assemble, compile, generate, interpret and compute.

process control: descriptive of systems in which controls are used for automatic regulation of operations or processes.

process I/O: input and output operations directly associated with a process, contrasted with I/O operations not associated with the process. For example, in a process control system, analog and digital inputs and outputs would be considered process I/O, whereas inputs and outputs to bulk storage would not be process I/O.

process variable: (1) in a control loop, the variable being controlled to the setpoint; (2) any parameter within the process that is of interest from an operations or control standpoint.

program: a set of computer routines used to solve a problem. The instructions are executed in the order they are written.

Programmable Read Only Memory (PROM): a nonvolatile (data will not be lost when power is turuned off) memory that is used to store unchanging information such as programs. The computer system cannot write information to a PROM. PROMs are loaded at the factory and can only be changed using special equipment.

proportional action: a controller mode that contributes an output proportional to the error.

proportional band: the range of the controlled variable that corresponds to the full range of the final control element.

protocol: a set of rules governing communications between computers that ensures messages are correctly sent, received and understood.

Random Access Memory (RAM): the portion of computer memory that can be both written to and read from.

286 PROCESS CONTROL COMPUTER SYSTEMS

rate time: for a linearly changing input to a proportional plus derivative mode controller, the time interval by which derivative action advances the effect of proportional action.

ratio control: control in which a secondary input to a process is regulated to maintain a preset ratio between the secondary input and an unregulated primary input.

real time: (1) actual time during which a physical process transpires; (2) performance of a computation during the actual time that the related physical process transpires in order that the results of the computations can be used in guiding the physical process.

reliability: the probability of successful operation of a system or component.

remote control: electronic control of valves and other operations located away from the area control center.

remote/local (R/L): refers to a switch setting at local panel controls. The switch in local position means that control can be exercised at the local panel only, locking out control from a remote location. The switch in remote position means that control is exercised from a remote location and local controls are locked out.

repeats per minute: controller integral mode adjustment units. The inverse of integral time.

report: a printed copy of process information and calculated values.

report generator: a computer program designed to be used with very little training. They are designed to allow the user to retrieve information from files in various report formats.

reset windup: in a controller containing integral action, the saturation of the controller output at a high or low limit due to integration of a sustained deviation of the controlled variable from the setpoint.

response time: the time a system takes to react to an input signal or command. Examples are the time to show a requested display or the time to shut down a pump on receiving an alarm.

reverse-acting controller: a controller in which the value of the output signal decreases as the value of the input (measured variable) increases.

robustness: the degree to which a control system can still function with some failed components (i.e., in a degraded control mode).

RPG (report program generator): a programming language used on some small business computers. It is usually considered a high-level language.

SCADA: *see* supervisory control and data acquisition.

screen refresh: a hardware function that maintains an image on a VDT screen by the continuous generation of a composite video signal from data stored in the memory of a VDT controller.

screen update: a software function that periodically replaces the dynamic data portion of a display with current real-time data.

self-configuring: a method of programming that eliminates the need to reassemble or recompile programs after a change in configuration by the dynamic use of parameters external to the program that define the particular configuration.

sequence control program: a high-level program whose primary function is to cause a sequence of events to happen based on current process requirements or operator requests.

sequential control: several valves, gates or equipment operations are caused to operate simultaneously or in a certain time sequence.

serial transmission: a mode of transmission whereby bits of a data word are sent sequentially (starting with either the most or least significant bit) over a single communications channel.

setpoint: in a control loop, the desired value of the process variable being controlled.

signal: (1) the event or phenomenon that conveys data from one point to another; (2) a time-dependent value attached to a physical phenomenon and conveying data.

simulation: the representation of certain features of the behavior of a physical or abstract system by the behavior of another system, for example, the representation of physical phenomena by operations performed by a computer or the representation of operations of a computer by those of another computer.

software: (1) a set of programs, procedures, rules and associated documentation concerned with the operation of a computer system, e.g., compilers, library routines and manuals; (2) a program package containing instructions for the computer hardware.

software subsystems: major segments of the software that perform a unique, identifiable function. This includes such subsystems as operating system, logging, scanning, graphic displays, alarming, DDC and each separate control function.

source language: the code a programmer produces in a high- or low-level language before it is translated into object or machine language.

steady-state: a characteristic of a condition, such as value, rate, periodicity or amplitude, exhibiting only negligible change over an arbitrary period of time. It may describe a condition in which some characteristics are static, others dynamic.

supervisory control: (1) a control technique in which a digital computer is used to determine and fix setpoints for conventional analog controllers (used to distinguish from direct digital control); (2) a high-level program whose primary function is to oversee an ongoing process and alter the general parameters of a control strategy based on mathematical relationships.

supervisory control and data acquisition (SCADA): a digital control system that can setpoint analog controllers and gather and store process data on-line.

synchronous transmission: (1) transmission in which the sending and receiving instruments are operating continuously at substantially the same frequency and are maintained, by means of correction, in a desired phase relationship; (2) a mode of data transmission whereby the message is sent in a continuous bit string.

system management: systematic direction of people to bring about the desired results of using the control system components in the most cost-effective manner.

system regeneration: the complete reconfiguration of a computer's software in main memory and mass memory. This function will typically require taking the computer control subsystem off-line.

system software: general computer programs that are of use in many applications. Examples are the operating system, report package, editor and compiler (*see* application software).

table-driven: a programming method whereby the paths of execution through a computer program are controlled by the contents of a data table containing logical or numerical data.

telecommunication: the use of leased telephone lines for sending informative data and performing operational controls at great distances.

telemetering: the transmission of a measurement over long distances, usually by electromagnetic means.

telemetry system: an audio tone system that controls and monitors water flow, hydraulic variables and alarms over a telephone line.

throttling control: control that directs a final control element to intermediate points within its operating range; distinguished from on/off control.

time constant: the time required for the output of a single capacity element to change 63.2% of the amount of total response when a step change is made in its input.

time slicing: a technique for allocating CPU time between multiple programs in a multiprogramming environment. This technique allocates a given segment of time to each task in a round-robin fashion without regard to task priority or importance.

top-down design: an organization of design where plans are made for modules arranged in a pyramid, starting at the top-most module, which is the most generalized concept of the control system.

transducer: an element or device that receives information in the form of one physical quantity and converts it for transmission, usually in analog form. This is a general definition and applies to specific classes of devices such as primary element, signal transducer and transmitter.

transfer function: a mathematical, graphical or tabular statement of the influence which a system or element has on a signal or action compared at input and output terminals.

transmitter: a transducer that responds to a measured variable by a sensing element, and converts it to a standardized transmission signal, which is a function only of the measured variable.

turnkey vendor: one who provides a complete system, including the computer, software, training and installation.

uninterruptible power supply (UPS): a power supply having backup battery storage. A UPS is used to ensure operation of critical computer equipment during power failures.

video display terminal (VDT): (1) an electronic vacuum tube containing a screen on which information may be stored for visible display by means of a multigrid modulated beam of electrons from the thermionic emitter; (2) loosely, a computer terminal using a cathode ray tube as a display device.

watchdog timer: an electronic internal timer that will generate a priority interrupt unless periodically recycled by a computer. It is used to detect program stall or hardware failure conditions.

word: (1) a character string or a bit string considered as an entity; (2) a group of binary digits treated as one unit of information and stored in a single memory location.

INDEX

acceptance, computer control system   133,259
active line power   99
ADA   31,76
air conditioning   65
alarm
   handling   105
   reporting   106,107
   summary   106,107
algorithms   31,51,77,96,189,190,213, 214,223
alphanumeric control displays   198
   example   200
alternatives, evaluation of   157
alternatives, identification of   155
analog
   control backup   3,6,42,49,56,121, 122,126,249
   controller   25,183
   input   28
   limit check   106
   process I/O   28,48,50,218
   signal conversion   28,48,64
   standards   48,55
   -to-digital (A/D) converter   28
   wiring   92
analytical instrumentation   63
annual cost comparison   158
application engineering   114–118,126
application software   72–76,79–85

as-built documentation   223
assembly language   71,73
attitude   242
authority   140,141,209,256
automatic control systems   96
automation   22,93
availability   187,192
   See also operational availability demonstration
background   104
backup control panels
   See analog backup
bar charts   198
BASIC   31,76
batch control   23,129
batch programming   74
block diagram   180
brainstorming techniques   149,155
bubbler level transmitter   61
budgets   136

calculations   215
capital cost avoidance   89
capital cost estimating   154
carrier current   42
cash flow   164
central analog   90
central computer   80
central digital control systems   49,90
centralization   23,101

291

292   PROCESS CONTROL COMPUTER SYSTEMS

central processing unit (CPU)   27,78
change control   266,270
changes   231
checkout   8,132
chemical use reduction   96
chip   70
chromatography   63
COBOL   31
color display terminal   105
compiled language   72,73
computer operator   34
computers, dual   232
computer structure   43
consistent plant operation   104
console operator   46
construction   239
consultants   16
contingency   187,190
   control   141
continuous control   24
control
   devices   43,64,111
   diagram   185,186,189
   engineer   82
   language   73,78,80
   loop   183
   manual   90
   modes   102,103
   modules   115
   panels   42,49
   philosophy   9,121,185
   programmer   35
   regulatory   24,77,80
   room   65,265
   sequential   23,56,77,80
   strategy   74,80,83,96,183
   strategy design checklist   184
   system analyst   35
   system design guide (CSDG)   179
   system engineer (CSE)   20,34
   system manuals   14,265,272
controllability   183
controlled (directive)
      management   256
corporate model   20
correspondence   209

cost reduction   111
crisis management   256
critical backup functions   121
custom modifications   125
cycle time   27

data
   acquisition   10,77,80
   base   22,77,78,80,83,193
   base generator   73,78,80
   base management system   32
   highway   49,52,82
   logger   26,53
   processing   3,19,20,27
debug   78
degraded control   190,192
delivery   215
density measurement   62
design   167–206
   benefits   178
   compartmental   182
   coordination   127,130
   engineering   171,174
   functional   176
   interdisciplinary   123
   lateral   173
   of algorithms   189
   of controls   183
   of data handling   193
   of OPI   197
   options   174,175
   top-down   84,167,168,174
   user   170,174
   vendor   173,174
designer   35
diagnostic   78
digital control benefits   87–112
   capital cost avoidance   89
   chemical usage   96
   energy   98
   intangibles   102
   labor   93
   operational data   105
   power factor   99
   product consistency   101
   throughput   156

INDEX

wiring   92
digital process I/O   48,50,218
digital technology   101
digitally emulated analog control (DEAC)   12,187
direct ditial control (DDC)   187
discrete control   24
discrete wiring   92
disk   47,50,76,85
diskette   47
dispatching systems   120
display generator   75,78,80
distributed analog   90
distributed control   49,52,77,79,80, 90,101,118
distributed controller   27
documentation   8,84,85,130,223

economic evaluation   154,157
energy use savings   98
engineer   169
engineering design   171,174
engineering fees   126
engineering units   76
equipment life   104
equipment out-of-service display   106,108
ergonomics   197
error   184
evaluation criteria   152
expansion   231
explicit design   176

factory testing   83,215,222
fail-safe mode   105
failure reports   224,226
feedback control   24,54,96
feedforward control   54,96
field changes   131
field mounted equipment   93
field testing   83,222
filter   76
fixed capacitors   100
flexibility   104
float level transmitter   61
floppy disk   47

flow chart   70,71,190,191
flow measurement   61
flow, split   184
FORTRAN   31,72,73,76
freeze dates   168
fuel, power and chemicals   156
functional design   176
future of microtechnology   89

graphics   198,199,209,214
grounding, computer system   66

hardware   3,26-29,41-68
high-level language   71,74,79,81,82, 84,86
historical data   106,108,193
HOA   55
human factors engineering   197
hybrid control systems   25,53,88,90

implementation   7,16,117,207-228
   plan   162
indicator   55
in-house implementation   208
inspection   217
installation   217
instruction set   27
instrumentation   43,63,111,218
   engineer   35
interactive data entry   173,198, 199,201
interface   123,217
interlocks   53
interrupt   30,71,74
interview techniques   151

keyboard   46,200
key operator   34
key person
   See system manager

lab data entry   106
lateral design   173
level measurement   60
LISP   31
load following   96

load shedding 99
local area network (LAN)
  See data highway
logic control 23
loop drawing review 212

machine code 71,73
magnetic flow meter 62
magnetic tape 47
maintenance
  preventive 231
  report 106,108
  reporting 108
  users 264
management 8,84,85,253-273
  approaches 256,266
  commitment 109,253
  delegation 254,257
  functions 254,263
  information 105
  responsibility 257
manual backup panels
  See analog backup
mass storage
  See disk
memory 28,72
menu selection 46
messages 215
microcomputer 44,50,59,79,81,
    110,126
minicomputer 45,79
momentary contact 28,55
multiplexer 26,28,42,49,53,58,77,88
multiprogramming 30

needs and objectives 151
negotiated bid 130,133
noneconomic evaluations 160
numerical control 23

O&M manuals 14,141,265,272
office automation 21
on-line data 193,195
open channel flow 61
open loop control 97
operating cost estimating 154

operating procedures 141
  reports 75,78,80,106,107,209
  savings 154
  system 30,72,73,76,80
operational availability demonstration
    (OAD) 83,204,223,224
operational changes 106
operational data reporting 105
operational labor 156
  avoidance 93
operational takeover
  See acceptance
operation by exception 93
operations apprentice 94
operations coordination 140
operator
  fears 139,260
  interface 77,80,105,193,197,211
  morale 261,270
  motivation 260
  overloading 120
  qualifications 94,130,137
  responsibilities 262,265
optimization 8,268
optimizing digital control (ODC) 12,
    33,187,188
organizational structure 259
orifice plate 62
owner 169

Panels 212,218
  backup 249
  control 42,49,53,56,121,122,
    126,249
Parshall flume 61
PASCAL 31,73,76
performance standards 185,187
peripheral 42,46,78,80
pH measurement 63
phone records 210
PID controller 24,44,55,77
planning 6,15,145-166
  procedures 147
  reasons for 146
  talent 15,119
PL/1 31

pneumatic control 54
power
   conditioning 66
   demand 50,98
   factor control savings 99
prequalification 130,134
present worth comparison 159
pressure measurement 60
printed circuit card 70
printer 46,51,52
process
   efficiency 102
   engineers 263
   input/output 28,48,50,218
   and instrumentation (P&IDs)
      diagram 188,189
   performance 92
   spare capacity 92
   variable 184
process control
   applications 23
   computer 43,49
   decision-making 101
   language (PCL) 32,212,213
   software 69,71
procurement 37,82
production consistency 101
productivity 102
product throughput 156
programmable controller 81
programmable logic controllers
      (PLC) 24,34,45
programmable nature of digital
      systems 102
programmer 34,264
programming languages 30,31
project
   milestones 163
   schedules 164
   size 123,128,258,259,262
   task force 150
proportional control 54
proprietary processes 117
protocol 53,59
punchlist
   factory test 217

   field test 223
   inspection 219
   OAD 225
pyrometer 62

range 193
reactive line power 99
read-only-memory (ROM) 51
real-time
   software 71,72,76
   information 101
   monitor 76
redundancy 180
regulatory control 24,77,80
reliability 103,121,187,192
remotely controlled facilities
   See dispatching systems
remote multiplexers (remote terminal
      units) 88
report generator 75,78,80,106,
      107,209
resistance temperature device
      (RTD) 62
response time 187
return on investment 159
risk analysis 162
robotics 23
robustness 190
rotation, staff 236

schedule 208
sensors (sensed variables) 106,183
sequence control 23,56,77,80
setpoint 24,25,54,184
shop drawings 131,210
simulation 192
site planning 65
software 29,69–86,105,204
   review 212
softwire 51
sonic flow meter 62
specialization, staff 236
specification 82,83,202,204
spectrum analysis 63
staff acquisition 135
staff focus 109

staffing   8,13,83,84,165,229-252
storage, backup   233,249
strain gauge   60
strategies
   See algorithms
submittals
   See shop drawings
success steps   5
summary display   200
supervisory control (SCADA)   10,187
suspended solids   62
switch, rotary   55
system
   analyst   34
   engineering   *See* application engineering
   integrator   36
   manager   4,8,109,136,254,256, 258-261
   operational description   126
   procedures   109
   software   74,75,76,78,79,81,82
   supplier   36
   theory   192

telemetry
   duplex   57
   equipment   42,46,56
   leased lines   57
   microwave   58
   protocol   59
   scanner   58
   telephone lines 3000   56
   telephone systems   56
   tone equipment   58
   UHF radio   57,59
temperature measurement   62
testing   8,219
thermistor   62

time-slicing   30
training   8,134,139,219,230
   maintenance   221
   on-the-job   230
   process   220
   software   220
   system management   221
transmittal   210
trend report   87,106,108,200
tuning   215,269
turbine flow meters   62
turnkey system   5

ultrasonic level   61
uninterruptible power system (UPS)   50,66
unions   134,136,260,265
updates, software   232
use of computer   104
user   169
   design   170,174
   participation   209
"user friendly" software   87
utility   78

valves   64
vendor   76,82,83,84,169
   design   126,173,174
   implementation   118
venturi   62
video display terminal (VDT)   77,82, 103,172

weighting system   152
weir   61
wiring
   costs   92,93
   distances   93
   interface   194
word length   27

12/15

UCC South Charleston, WV   770
TS156.8 .S73
/Process control comp

1023950